Advances in Analytics and Data Science

Volume 1

Series Editors
Ramesh Sharda
Oklahoma State University, Stillwater, OK, USA

Hsinchun Chen
University of Arizona, Tucson, AZ, USA

More information about this series at http://www.springer.com/series/15876

Murugan Anandarajan • Teresa D. Harrison
Editors

Aligning Business Strategies and Analytics

Bridging Between Theory and Practice

Springer

Editors
Murugan Anandarajan
Drexel University
Philadelphia, PA, USA

Teresa D. Harrison
Drexel University
Philadelphia, PA, USA

ISSN 2522-0233 ISSN 2522-0241 (electronic)
Advances in Analytics and Data Science
ISBN 978-3-030-06634-5 ISBN 978-3-319-93299-6 (eBook)
https://doi.org/10.1007/978-3-319-93299-6

This Springer imprint is published by the registered company Springer Nature Switzerland AG
The registered company address is: Gewerbestrasse 11, 6330 Cham, Switzerland

Preface

As a co-op institution, Drexel University has long-since recognized the need for better linkages between academic/theoretical content and on the job training. At LeBow, we have embraced this linkage through the sponsorship of an annual conference called Bridging Practice and Theory Summit (BPTS). Historically, BPTS invited keynote speakers with key insight into current challenges faced by the business community and the content primarily focused on a traditional academic conference format of concurrent sessions with faculty presenting their findings on innovative pedagogy. We begin to recognize the need to bring industry into the summit as a full participant. Our conference focused on two themes: (i) the need to bring current business issues into the classroom and relate these issues to current theory without decreasing instructional rigor; (ii) the need to reinforce the importance of increased dialogue between practitioners and academics.

BPTS seeks to foster this dialogue with a strong belief that these interactions between practicing managers and researching academics are important in increasing both the relevance of pedagogy and research vital for enhanced learning and bidirectional problem solving. That is, theory should inform practice and vice versa.

From this mission, BPTS 2017 was developed with a focus on business analytics, given the growing need for strategic and systematic decision making, informed by data and analytics. While we can all likely agree that fostering these relationships would be beneficial, developing the mechanism for such collaboration is not inherently structured into business or academic practices. Indeed, conferences such as this are sorely missing in the current landscape. Academic conferences often focus on specific academic disciplines with a lens toward research while overlooking the pedagogical and applied perspectives, while practitioner conferences lean toward tools and techniques and do not address a need to bring practice to the classroom.

This conference uniquely highlights the perspective that we should more actively involve the business community in this conversation to seek more integrated solutions and potentially create more joint collaborations between businesses and universities.

We therefore structured BPTS 2017 such that all academic sessions would be co-led by an academic and a practitioner in the field. Faculty and industry experts developed the topics, sessions, and presentations jointly, promoting a sustained engagement in thinking about these important topics. The chapters you see here are the culmination of that collaboration and therefore, collectively, provide a unique and novel perspective on the ways that business analytics should grow as a field to advance business practice.

We would like to thank Raji Sivaraman, PMI; Chris Radvansky, PRO Unlimited; Samir Shah, Drexel University; David Kurz, Drexel University; Gerard Olson, Villanova University; and Steven Pyser, Rutgers University, for their valuable reviews of the chapters in this volume.

Philadelphia, PA, USA

Murugan Anandarajan
Teresa D. Harrison

Contents

Contributors

Murugan Anandarajan Drexel University, Philadelphia, PA, USA

Bay Arinze Drexel University, Philadelphia, PA, USA

Atif Ashraf Sogeti, New York, NY, USA

Greg Baldini Nihilent Technologies, Inc, Philadelphia, USA

Suresh Chandran Drexel University, Philadelphia, PA, USA

Amit Devpura AstraZeneca, Cambridge, UK

Lawrence Duke Drexel University, Philadelphia, PA, USA

Alexander Gochtovtt Epsilon, Philadelphia, USA

Teresa D. Harrison Drexel University, Philadelphia, PA, USA

Mark Holder AstraZeneca, Cambridge, UK

Rahul Kasat DuPont, Wilmington, DE, USA

Russ Kliman SEI Investments Company, Oaks, PA, USA

Deanne Larson Larson & Associates, LLC, Seattle, WA, USA

Anthony Lee A4P Inc., Philadelphia, PA, USA

Sueanne McKniff WellSpan Health, Harrisburg, PA, USA

Rubén A. Mendoza Saint Joseph's University, Philadelphia, PA, USA

Gloria Phillips-Wren Loyola University Maryland, Baltimore, MD, USA

Ali A. Poorani Hospitality Associates for Research & Training (HART), Hospitality Business Management, Alfred Lerner College of Business and Economics, University of Delaware, Newark, DE, USA

Bill Powell ARI, Mount Laurel, NJ, USA

Elven Riley Seton Hall University, South Orange, NJ, USA

Mark Schild Seton Hall University, South Orange, NJ, USA

Samir Shah Drexel University, Philadelphia, USA

William A. Sullivan Courtyard Newark at the University of Delaware, Newark, DE, USA

About the Authors

Murugan Anandarajan is a Professor of MIS at Drexel University. His current research interests lie in the intersections of areas Crime, IoT, and Analytics. His work has been published in journals such as *Decision Sciences*, *Journal of MIS*, and *Journal of International Business Studies*. He coauthored eight books, including *Internet and Workplace Transformation* (2006) and its follow-up volume, *The Internet of People, Things and Services* (2018). He has been awarded over $2.5 million in research grants from various government agencies including the National Science Foundation, US Department of Justice, National Institute of Justice, and the State of PA.

Bay Arinze is a Professor of MIS in the Decision Sciences Department at Drexel LeBow College of Business and a faculty member for Corporate and Executive Education. His current research focuses on data mining, text mining, and predictive analytics. Dr. Arinze has published numerous articles in several prestigious journals. He has received many private and public research grants and has consulted widely in information technology with clients including the US Marine Corps. Dr. Arinze received his MS and PhD degrees in Information Systems from the London School of Economics.

Atif Ashraf is a Business Intelligence and Analytics Consultant with a focus in the Pharmaceutical and Consumer Product industries. His experience includes data visualization in custom dashboards using Tableau, architect of AWS environments, and analyzing data using the open-source software platform R. He is a certified Amazon Web Service Solutions Architect Associate and a Certified Scrum Master professional. His focus is on customer insight analysis, marketing optimization, and data visualization. He holds an MS in Business Analytics from Drexel University and a BA in Economics and Communication from Rutgers University.

Greg Baldini is a Solutions Architect with Nihilent Technologies, Inc. He leads the data practice for the east coast and business analyst practice nationally. His experience lies in delivering effective analytical solutions starting from the ideation phases and following through to rollout, training, and adoption. His goal is to deliver on the promise of analytics: make better decisions through data. An expert in data modeling and engineering, systems architecture, and the Microsoft Data Platform, he has delivered solutions in many spaces including manufacturing, healthcare, fraud analysis, marketing pipelines, and sales.

Suresh Chandran is an Associate Clinical Professor of Management at Drexel's LeBow College of Business. Suresh's interests are in the area of Technology Management, Innovation, and Strategy, and he has over 20 years of experience providing consulting services to national and international companies. He received his PhD from Vanderbilt University.

Amit Devpura heads the Supply Chain Analytics group at AstraZeneca and is responsible for improving supply chain performance metrics through advanced analytical techniques. He has built a strong Analytics group that prides itself in turning data into information and steering the organization on the Analytics maturity curve. Amit has vast industry experience where he has led Analytics function in technology, pharmaceutical, and apparel and footwear industries. Amit holds a PhD in Industrial Engineering from Arizona State University and an MBA from Wharton Business School at University of Pennsylvania.

Lawrence Duke is a Clinical Professor and also serves as a faculty advisor to the MS Marketing program at Drexel University LeBow College of Business. His research focuses on marketing ethics, quality, digital, and global marketing and finance. Prior to becoming a business professor, he was a senior executive in international finance and marketing. He has 20 years of experience in the public and private sectors, including positions at Citigroup, Nomura, State Street, and the International Monetary Fund (IMF), and as a US naval officer.

Alexander Gochtovtt is a Senior Vice President and General Manager at Epsilon Data Management, a subsidiary of Alliance Data Systems, where he focuses on providing database marketing and digital transformation services to Fortune 100 clients in Life Sciences. His experience in data management and emerging technologies comes from a variety of roles at consulting firms, startups, and digital ad companies. Most recently, Alex has been focusing on privacy and blockchain distributed ledger technologies. He received a degree in Information Systems from the University of Scranton.

Teresa D. Harrison is Associate Professor of Economics, Associate Dean for Academic Affairs, and Director of the Center for Teaching Excellence at Drexel University's LeBow College of Business. She is an applied micro-economist, with interest, research, and publications in nonprofit organizational behaviors, including location decisions, entry, exit, mergers, and growth. Her work has been published in journals such as *Financial Review* and *Economic Inquiry*.

Mark Holder is the Executive Director of Regional Supply North America at AstraZeneca. Mark is accountable for the leadership of the North American Demand Team responsible for the interface between Operations and the North American Commercial Organization ensuring alignment between Commercial, GPPS, and Operations strategies to maximize AZ supply chain value outcomes. Mark began his career as a commissioned officer in the United States Navy followed by positions in high-tech manufacturing and supply chain consulting before joining AstraZeneca. Mark holds a bachelor degree in Commerce and Engineering and an MBA from Drexel University.

Rahul Kasat is a Global Business Leader at DuPont. In his current role, he is responsible for launching new business in the area of 3D printing for DuPont's performance materials business. He has held several positions in DuPont in the areas of technology, strategy, marketing, and new business development. Rahul is a co-inventor on 5 granted patents and co-author on 11 peer-reviewed articles and 1 book chapter. He received his PhD in Chemical Engineering from Purdue University and his MBA from Drexel University.

Russ Kliman is an Innovator and Design Leader for Challenging Mental Models at SEI as well as leader of SEI's Global Innovation Program. Russ' focus is to lead various programs which have direct strategic impact across all of SEI's markets globally. At various stages of his tenure with SEI, he has led various strategic initiatives from new business marketing and business development to product design and technology leadership. Russ holds a Bachelor of Science degree in Information Systems Design from Drexel University.

Deanne Larson is an analytics practitioner and principle faculty at City University of Seattle. Dr. Larson's research has focused on qualitative studies in developing an enterprise data strategy and identifying best practices in agile business analytics.

Anthony Lee is Principal in performance solution with Propensity4. Anthony's career began in supply chain and demand planning at Alcoa, and later in commercial forecasting and business planning, becoming head of US Forecasting Group at AstraZeneca. As Global Commercial Capability Development Director he standardized the forecasting stewardship and accuracy methodology for AZ's Oncology franchise. Recent highlights at Propensity4 involve integrating Big Data and performance to operationalize strategic forecasts for pre-launch products. He received his BSc, MEng, and PhD in Computer and Systems Engineering at Rensselaer Polytechnic Institute.

Sueanne McKniff is the Program Manager for Community Connect at WellSpan Health. Her clinical experience and interest in process efficiency resulted from 18 years of nursing in WellSpan's Emergency Department and Trauma

Resuscitation Area. She is Board Certified in informatics and received an Executive MBA from Loyola University of Maryland. Her research interests and publications are in workflow analysis, process efficiency, and the application of information technology in the healthcare arena with a focus on patient safety.

Rubén A. Mendoza is an Assistant Professor in the Decision & Systems Sciences Department at the E. K. Haub School of Business, Saint Joseph's University. He holds a bachelor's degree in Electrical Engineering from the City College of New York (CCNY), an MBA, and a master's degree in Information Science from the University of Michigan. He completed his PhD in Management Information Systems at Rensselaer Polytechnic Institute. His research is in diffusion of innovations in the area of Business Intelligence.

Gloria Phillips-Wren is Professor and Chair of Information Systems, Law and Operations at Loyola University Maryland. She is co-editor-in-chief of *Intelligent Decision Technologies*, associate editor of *Journal of Decision Support*, and highly active in other aspect of the knowledge and decision sciences field. Her research interests and publications are in decision support, analytics, business intelligence, healthcare IT, and strategic uses of technologies in organizations. She received her PhD from the University of Maryland Baltimore County and holds MS and MBA degrees.

Ali A. Poorani is an Associate Professor at the University of Delaware with expertise in Leadership and Entrepreneurship. He is the Director of Hospitality Associates for Research and Training, offering research and training programs for the hospitality and healthcare industries. He is the Director of the University of Delaware's Patient Experience Academy, a partnership between Lerner College of Business and Economics and Christiana Healthcare System. Dr. Poorani received his PhD in Leadership and Human Behavior from USIU and his MBA from University of Central Oklahoma. He has published several scholarly articles.

Bill Powell is the Director of enterprise architecture for ARI (Automotive Resources International). Bill's area of research and expertise is focused on the application of analytics within the fleet management and automotive markets to drive efficiencies and to create innovative products and services. Bill's current efforts are focused on the coupling of in-memory systems with IoT (Internet of Things) data.

Elven Riley is Director of Securities Analysis and Trading at Seton Hall University Stillman School of Business. Continuing decades of investment banking technology practice, Elven's research remains in disruption and behavioral change within the financial services ecosystem.

Mark Schild is Assistant Dean of the Stillman School of Business, at Seton Hall University. In addition, Mark teaches finance and investment related courses leveraging his 30-plus years' experience owning and operating a broker-dealer and investment advisory firm to transfer concepts into practice.

Samir Shah is a Clinical Professor in Management Information Systems within LeBow College of Business at Drexel University. Over the past 20 years Samir's vision to connect technologies to solve business problems has been strengthened through his progressive experience in various dimensions of information technology at government, private sector, and academia. Shah's research interest is in the areas of IT project management, agile software development, and design thinking in IT projects.

William A. Sullivan is the Managing Director of the Courtyard Newark at the University of Delaware and manages the "experiential training" for students in the University of Delaware's Hospitality Business Management program. He is employed by the Shaner Hotel Group as the on-site manager for the University of Delaware Blue Hen LLC hotel. Bill is also an adjunct faculty instructor for the UD Program at Widener University. He holds a BS in Business Administration and MS in Hospitality Business Management from the University of Delaware.

List of Figures

List of Tables

Chapter 1
Aligning Business Strategies and Analytics: Bridging Between Theory and Practice

Murugan Anandarajan and Teresa D. Harrison

Abstract In this chapter, we discuss the current gap between academic training and the needs within the business community, the potential for this gap to widen, and the role joint work between academic and industry experts can play in bridging this gap. We highlight the particular case of business analytics, calling attention to the current business landscape and the need for strong training of future employees, grounded in both rigorous theoretical background and links to the practical applications. The chapter concludes by emphasizing the particular contributions of each chapter and making a case for this type of work to be among the first of many steps in creating more meaningful dialogue between higher education and business practitioners.

Keywords Alignment · Business analytics · Theory · Practice · Tools

Introduction

The gap between theory and practice is widely documented and debated in the business literature (Aram & Salipante, 2003; Argyris & Schon, 1974; Wren, Halbesleben, & Buckley, 2007). In general terms, the theory-practice gap can be defined as the discrepancy between what students acquire through the theoretical classroom lectures and what they experience in the workplace (Ajani & Moez, 2011). There are two main issues causing the theory-practice gap. First, some of the theory is too idealistic and impractical. Second, even where the theory is practical and beneficial to the organization, some practitioners do not act on it, possibly due to ignorance and the rigid system in which they work or because they choose to ignore it.

While the gap between theoretical content and practical knowledge has existed for some time, the need to facilitate the connections between the two has increased

M. Anandarajan (✉) · T. D. Harrison
Drexel University, Philadelphia, PA, USA
e-mail: ma33@drexel.edu; tharrison@drexel.edu

M. Anandarajan, T. D. Harrison (eds.), *Aligning Business Strategies and Analytics*, Advances in Analytics and Data Science 1,
https://doi.org/10.1007/978-3-319-93299-6_1

dramatically over the last decade. Business decisions are becoming increasingly complex and interdisciplinary, calling on knowledge across many business and functional areas. This requires better grounding in theoretical content but also increased knowledge across many domains.

Within academic institutions, the business curricula are often frontloaded with theories on the grounds that students need to learn the theories before entering the workforce. This is based on two implicit assumptions. First, these theories are applicable in the real-life context for informing good practice. Second, formal theories constitute the best grounding for the valid knowledge of practice. These theories are often taught in a didactic manner as decontextualized abstractions and generalizations, to be applied later by students in their careers. This approach has historically and traditionally been the preferred dissemination of the relevant theories under the premise that the particular business and work-related applications will evolve dramatically and (perhaps) rapidly over a student's career and thus the theories provide the proper basis for learning new techniques, technologies, and methodologies.

Unfortunately, this approach leads students and working professionals to often perceive a disjunction between the abstract form of these theories in texts and their lived form in practice. As a consequence, many students find these theories to be magniloquent and irrelevant to practice. Unless we are able to help students make sense of the link between theories and practice, students will face problems in bridging the gap between the two.

Moreover, to the extent that organizations, business executives, and hiring managers interpret this gap as implying that universities are not properly preparing students for the current workforce, the importance and relevance of higher education will be diminished. There are viable reasons why companies may view this gap as widening. The pace of business decisions and the velocity of changes to the business environment have increased rapidly over the last decade. These types of changes filter down to business' needs for job preparedness, which has increased tremendously (Ahmad & Pesch, 2017). Organizations do not have the luxury of time for new hires to "learn on the job." They require individuals who can become immersed in the business problems at hand quickly and develop relevant and timely solutions. It is therefore no surprise that organizations may perceive that students are not as equipped for this newer business environment. Such perceptions are a main catalyst for corporations significantly increasing their expenditures on workforce training.

Ultimately, we need greater synthesis of theory and practice if we are to prepare thoughtful practitioners (Raelin, 2007, p. 495). In business education, even at the undergraduate level, business schools are doing the work of preparing practitioners. To prepare someone implies providing a link, building a bridge, or making a connection. Preparing thoughtful practitioners requires moving away from the apparent dichotomy of theory and practice and moving toward the synergistic combination of the two. Thoughtful practice is informed by the complements of theory and practice. As we discuss further below, this volume in our view highlights this natural complementary and provides insights into why and how deeper connections between the academic and practitioner community will lead to more impactful learning, for students, academics, and practitioners, and ultimately stronger academic and business institutions.

Gap in Business Analytics

The discipline and study of business analytics are not immune to this theory-practice gap. Many organizations have become increasingly more dependent upon business analytics to obtain a competitive advantage. According to *Forbes* magazine, the analytics market will grow to $203 billion by the year 2020 (Press, 2017). However, despite the enormous corporate dependence on analytics, it is estimated that as few as one-third of new analytics projects produce a positive return-on-investment and that over two-thirds of all major analytics projects end in failure (Agarwal, 2017). In a recent study, the Gartner group noted that the major factors leading to analytics project failure are largely attributable to the lack of skills to deploy analytics-based tools and technology (Gartner, 2015).

Thus, although businesses across the world have invested billions of dollars in analytics in the last decade, organizations have found it difficult to harness the true value of analytics for long-term benefit, even though there is evidence analytics can transform industries and business practices. Aligning business strategies and analytics remains an elusive goal.

It is not unreasonable to posit that part of the failure can be attributed to the proficiency gap between knowledge and skills taught by academics and expertise and know-how requirements necessitated by organizations. It is also possible that techniques learned at institutions of higher learning become obsolete by the time the student joins the work force. As discussed above, we argue that this rapidly chang-

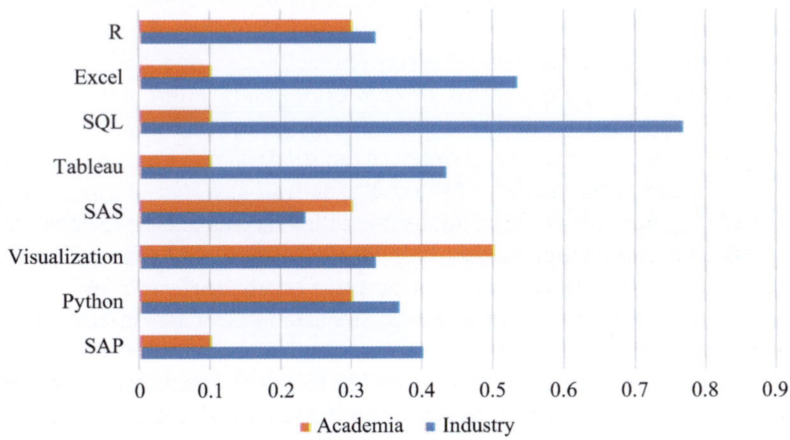

Fig. 1.1 Analytics tools: academia vs industry

ing environment necessitates the grounding to strong theory but also influences how both parties view the relative importance of current methods and techniques.

The authors examined the nature of the gap between skills taught by academics and those demanded by industry in terms of business analytics tools. The study found that there was a misalignment between academia and practice for most of the analytics tools. As can be seen in Fig. 1.1, academia places an emphasis on SAS that is unmatched by industry needs, whereas businesses have a need for professionals experienced with Excel, SQL, SAP, and Tableau that academia is not meeting.

With projections for the demand of skilled business analytics talent to triple over the next decade (Henke et al., 2016), organizations face challenges employing the required numbers of skilled professionals. This presents a two-pronged challenge. First, organizations confront the possibility of hiring business analysts with limited or no skills and second, organizations additionally face the prospect of having a shortfall of needed business analysts. This could be precarious for organizations in the current business environment where predictability, reliability, and efficiencies of analytics staff are paramount.

One potential way to reconcile academic preparation with industry need is to increase the conversations between academia and industry, so that the gap between them could be bridged more effectively. The chapters in this volume are the result of such a conversation. The BPTS 2017 conference organized sessions such that all academic sessions would be co-led by an academic and a practitioner in the field or practitioners who crossed the boundary. The chapters you see here are the culmination of that collaboration and therefore, collectively, provide a unique and novel perspective on the ways that business analytics should grow as a field to advance business practice.

The chapters are broken into three sections. In the first section, the chapters provide an insight into analytics in the finance, transportation, and biopharmaceutical industries. In the second section, the chapters discuss the importance of methodologies in analytics. In the final section, the chapters provide strategies by which alignment can be achieved between strategies and analytics. As highlighted below, the chapters also discuss pitfalls that both academicians and practitioners should be aware of that could make analytics less effective.

Part 1: Business Analytics in Practice

There are five chapters in this section. In Chap. 2, Kliman and Arinze examine how the delivery of financial advice and how cognitive computing can provide value for both the financial intermediary and the end consumer. For the intermediary, the study will assess how cognitive computing can augment and supercharge the expertise of the financial advisor, enabling the advisor to deliver improved advice. For the consumer, the study will assess how cognitive computing on its own (without a financial advisor) can deliver comprehensive and accurate advice comparable to that of a human advisor. The study will assess various aspects of cognitive computing, including but not limited to sentiment analysis, natural language processing, predictive analytics, and prescriptive analytics.

The chapter by Schild and Riley follows and discusses how technology is accelerating the growth of the financial advising industry at the same time that the wealth accumulated by older generations migrates slowly to younger generations. They argue that the preferred trust relationship will remain between human client and human advisor but will be heavily machine augmented with analytics and big data, delivered via the cloud.

In Chap. 4, Powell and Chandran use the case study of a leading fleet management company (ARI) and explore the application of advanced analytics to various facets of the industry and the company's experience in aligning analytics with its business strategy. Finally they outline the steps needed to implement a telematics and analytics strategy in organizations and the importance of bridging the gap between theory and practice.

Chapter 5 provides the context for analytics in practice within the biopharmaceutical industry. Holder, Lee, Devpura, and Chandran provide an insightful view of how the implementation of a supply chain blueprint model and value stream mapping has enabled tremendous cost savings at AstraZeneca. The importance of fostering a two-way dialogue between members of the business community and educators and introducing new programs like the Future Leaders program and Supply Chain Boards in bridging the gap between theory and practice through meaningful partnerships is also discussed.

Part 2: Methodological Issues in Business Analytics

In Chap. 6, Shah, Gochtovtt, and Balini offer real-world examples of how project management professionals tackle big data challenges in a rapidly evolving, data-rich environment. Simultaneously, project management establishes a bridge between business and academia as they both recognize the joint necessity to develop highly trained project managers to utilize the powerful and cutting edge analytical tools available to create value. This is followed by Larsen who explores the application of agile methodologies and principles to business analytics project delivery.

Phillips-Wren and McKniff discuss the operational benefits that can be gained by implementing real-time, big data analytics in a healthcare setting and the concomitant influence of organizational culture on adoption of the technology. They demonstrate these benefits by investigating patient-physician interactions in a large medical practice at WellSpan Health and compare the observed workflow with a modified one made possible with a big data, real-time analytics platform.

Chapter 9 by Poorani and Sullivan provide a case study on human capital analytics and investigate if such analytics add new outlooks beyond the usual metrics used by lodging enterprises. In addition, the case study provides measures that help management identify and address inefficiencies, as well as the productivity of its work force, with the goal of improving resource allocation.

Part 3: Aligning Strategies and Business Analytics

In Chap. 10, Mendoza reviews the opportunities and potential shortfall influencing the impact of business intelligence and analytics services for a company's internal use. He describes three strategies for providing these services internally and explores issues of importance in the shaping of current demand and of future offerings by Web-based providers. He concludes the section by discussing opportunities for the development of academic curricula and research that would offer better training to students, improve recruiting outcomes for organizations, and better address topics of current and strategic importance to the firm.

Duke and Ashraf in Chap. 11 draw attention to how new media marketing and analytics has fostered new insights about the customer journey, such as the creation of the loyalty loop and the need for alignment in marketing strategy. The implications for analytics education are also examined in the chapter with recommendations for curricula shifts and training as they relate to higher demand for and a shortage of qualified graduates.

In the final chapter, Kasat and Chandran discuss how data and analytics are playing a revolutionary role in strategy development in the chemical industry. This paper provides an overview of the challenges confronting the chemical industry and the opportunities to transform the industry by aligning data analytics and strategy. Using the case study of DuPont, they provide an example of how applying data and analytics to its precision agricultural technology increased yields and improved productivity.

Conclusion

It is important for academics and practitioners to recognize that for organizations to use data and analytics, it should be able to clearly articulate its purpose and then translate it into action throughout the organization. Similarly academia should be informed by practice about the analytics required by industry to be able to develop a relevant and rigorous curriculum. This volume uniquely highlights the perspective that academic institutions and practitioners should more actively collaborate to seek more integrated solutions. The collaboration seen in these chapters fosters this dialogue with a strong belief that these interactions between practicing managers and researching academics are important in increasing both the relevance of pedagogy and research and are vital for enhanced learning and bi-directional problem solving. We believe this volume emulates the types of collaborations and conversations that we hope to see develop and increase in the future.

References

Agarwal, S. (2017). Why big data projects fail and how to make 2017 different. *NetworkWorld*. Retrieved from https://www.networkworld.com/article/3170137/cloud-computing/why-big-data-projects-fail-and-how-to-make-2017-different.html

Ahmad, S., & Pesch, M. (2017). Essential work skills and readiness: Perceptions of employers, MBA students and undergraduates. *Academy of Educational Leadership Journal, 21*(1), 1–11.

Ajani, K., & Moez, S. (2011). Gap between knowledge and practice in nursing. *Procedia-Social and Behavioral Sciences, 15*, 3927–1931.

Aram, J. D., & Salipante Jr., P. F. (2003). Bridging scholarship in management: Epistemological reflections. *British Journal of Management, 14*, 189–205.

Argyris, C., & Schon, D. (1974). *Theory in practice: Increasing professional effectiveness*. San Francisco: Jossey-Bass.

Gartner (2015). Gartner survey highlights challenges to Hadoop Adoption. *Gartner*. Retrieved from https://www.gartner.com/newsroom/id/3051717

Henke, N., Bughin, J., Chui, M., Manyika, J., Saleh, T., Wiseman, B., et al. (2016). The age of analytics: Competing in a data-driven world. McKinsey Global Institute. Retrieved from https://www.mckinsey.com/~/media/McKinsey/Business%20Functions/McKinsey%20Analytics/Our%20Insights/The%20age%20of%20analytics%20Competing%20in%20a%20data%20driven%20world/MGI-The-Age-of-Analytics-Full-report.ashx

Press, G (2017). Six predictions for the $203 billion big data analytics Market. Forbes, January 20th. https://www.forbes.com/sites/gilpress/2017/01/20/6-predictions-for-the-203-billion-big-data-analytics-market/#7ff208242083

Raelin, J (2007). Toward an epistemology of practice. *Academy of Management Learning and Education, 6*, 495–519.

Wren, D. A., Halbesleben, J. R. R., & Buckley, M. R. (2007). The theory-application balance in management pedagogy: A longitudinal update. *Academy of Management Learning and Education, 6*(4), 484–492.

Part I
Business Analytics in Practice

Chapter 2
Cognitive Computing: Impacts on Financial Advice in Wealth Management

Russ Kliman and Bay Arinze

Abstract Cognitive computing is a form of problem-solving that incorporates machine learning, big data, data mining, natural language processing, machine vision, robotics, and other strands of artificial intelligence. Cognitive computing solutions can be used as sole or partial solutions to augment decision-making. The financial services industry is in a state of transformation, driven by the convergence of rapid changes in financial service technologies (fintech) – including cognitive computing, the digitization of the consumer, the emergence of younger investors (millennials), increased regulatory scrutiny (DOL regulation), and continued fee compression for products and services. Cognitive computing offers a disruptive opportunity in the financial services industry by not only empowering the financial intermediary but also by delivering increased engagement and value to the consumer.

This study examines how the use of cognitive computing to improve financial advice can provide value for the financial intermediary and the end consumer. For the intermediary, the study will assess how cognitive computing can augment and supercharge the expertise of the financial advisor, enabling the advisor to deliver improved advice. For the consumer, the study will assess how cognitive computing can deliver high-quality, accurate advice comparable to that of a human advisor.

Keywords Cognitive computing · Analytics · Big data · Robo-advisors · Wealth management · Financial advisers

---

R. Kliman
SEI Investments Company, Oaks, PA, USA
e-mail: rkliman@seic.com

B. Arinze (✉)
Drexel University, Philadelphia, PA, USA
e-mail: arinzeob@drexel.edu

M. Anandarajan, T. D. Harrison (eds.), *Aligning Business Strategies and Analytics*, Advances in Analytics and Data Science 1,
https://doi.org/10.1007/978-3-319-93299-6_2

Introduction to Cognitive Computing

Cognitive computing is a form of computer-based problem-solving that has already begun to transform companies and industries in profound ways. It incorporates four major strands, namely: machine learning, big data, data mining, and artificial intelligence. This new form of computing involves computer systems that can add to their own knowledge, utilize both structured and unstructured data, "reason" and infer, and frequently interact with their users in natural language.

The International Data Corporation (IDC) has identified cognitive computing as a major factor driving digital transformations in companies and changing the ways they function. IDC forecasts spending on cognitive systems to reach $31.3 billion by 2019, with a compound annual growth rate of 55%. Over 40% of the spending will be on software, with the second largest category being consulting and business services (IDC, 2016).

In regard to big data, cognitive systems typically leverage not only structured data but also unstructured data, such as free-form speech, email, click streams, government reports, research articles, social media, and opinion pieces and process the data to solve problems, reach conclusions, and assist workers in making decisions.

Cognitive computing also incorporates data mining, namely, the ability to identify patterns, causality, correlations, and actionable hypotheses in data. Such functions include descriptive, predictive, and prescriptive capabilities that help users gain insights into deep patterns in data and take more data-driven actions.

The road to cognitive computing, once traveled on by a few innovators and risk takers, is now crowded with small and large companies alike seeking to take advantage of the power and opportunities offered by these new systems. Why then have organizations begun to adopt cognitive computing systems? Here are a few reasons:

- They allow deeper and more fine-grained engagement with customers by leveraging structured and unstructured data. Organizations gain deeper insights into their customers' (or employees') behavior and mindsets through the use of such systems.
- Cognitive systems can also supercharge the expertise of workers in the breadth, depth, and quality of generated insights. Various cognitive computing "assistants" can collect and use vast knowledge from a wide variety of fields to assist human decision-makers.
- Next, cognitive computing enables organizations to radically transform their business processes, goods, and services by cognitively enabling them. This form of product differentiation serves not only to differentiate products but also to present customers with powerful new capabilities and services.
- The knowledge discovery and learning that are possible with cognitive systems are beyond both the reach of traditional modeling and programming and human cognitive abilities. Cognitive systems continually learn and increase their knowledge in ways that are well beyond the capabilities of human experts.

Cognitive computing is bringing value to financial management, and in particular, wealth management. The focus of a financial advisor, and thus the delivery of financial advice, is about the relationship with the client. It is through this relationship that the human advisor delivers value. In order for cognitive computing to deliver true value, it must enrich the relationship both economically and psychologically for client, the advisor, and the intermediary for which the advisor works.

Williams (2016) underlines that insisting that people "… tend not to trust things we don't understand or that threaten our jobs. Full adoption of cognitive-enabled DSS will have to overcome both of those hurdles to fulfill its true promise." When IBM's Watson described Toronto, Canada, as an American city (Reynolds, 2016), it showed the paradox of humans simultaneously being able to trust self-driving cars but being wary of systems that can make mistakes no human would make. Accenture's study (Shanks, Sinha, & Thomas, 2015) also showed that 84% of managers believed intelligent machines would make them more effective and their work more interesting. The study also drilled down to better understand the elements of trust: managers indicated that in order to trust cognitive systems, they needed to understand how the system worked and generated advice (61%), plus a proven track record (57%) and convincing explanations from the system (51%).

Financial intermediaries in wealth management must embrace cognitive computing in their efforts to deliver deeper advice, greater value, and at economic scale. To do so, they must recognize that the role of cognitive computing, within the context of the financial institution, is not to replace the human advisor but to amplify the human advisor's knowledge and capabilities, as well as to address the cognitive limitations of the human advisor. Advisors and the financial intermediary must also recognize that commoditization has occurred for elements that used to be part of the advice value chain. This includes risk assessment, asset allocation, portfolio construction and implementation, and reporting. This means they must make space in their work processes and thinking to allow the cognitive systems to assist them in these areas and not repeat prior analyses.

Cognitive systems therefore promise to unleash huge productivity gains via new and improved solutions to business problems. The future will see an oncoming wave of autonomous and semiautonomous systems that are able to respond independently to problems they are presented with.

The aim of this paper therefore is to describe and clarify how cognitive computing can provide value for both the financial intermediaries and end consumers.

Defining Cognitive Computing

Many understand cognitive computing from the performance of IBM Watson's victory over the best human players of the TV game Jeopardy. In the program, players, including Watson, had to listen to questions in English, reason through to an answer quickly, and provide an answer – also in natural language (also English).

IBM defines cognitive computing as: "an evolution of programmatic computing that enables a system to formulate responses on its own, rather than adhere to a prescribed set of responses." This differs from programmatic computer-based systems, where the range of possible responses from the system are predetermined. In cognitive systems, outputs are often unanticipated and outside a predetermined range.

Cognitive systems are designed to simulate human thought and reasoning using such techniques as data mining, natural language, and pattern recognition. These computer-based systems also handle ambiguity and tolerate unpredictability unlike programmatic systems that operate in predetermined ways. Watson's play in Jeopardy illustrated the major elements of cognitive computing: (a) sensing, (b) learning, (c) inferring or thinking, and (d) interacting (in natural language) – elements that are now being applied to many business problems and contexts.

Cognitive Computing: A Brief History

Cognitive computing has its roots in the artificial intelligence (AI) work of the 1960s and 1970s, when expert systems such as ELIZA were created to diagnose disease and techniques were developed for machine vision and natural language processing. Some see earlier roots from the late nineteenth century, when Charles Babbage proposed the "analytical engine" (Roe, 2014). John McCarthy coined the term "artificial intelligence" in 1955 and pointed to the potential for making intelligent machines. AI fell out favor in the 1980s and 1990s after the promises of the earlier two decades failed to be delivered, mainly due to a lack of processing power.

AI provides a major strand of cognitive computing, which is today's umbrella term that incorporates machine learning, natural language processing, and other capabilities described earlier in the paper. It is both a contemporary term and an expansion of the original scope of artificial intelligence by incorporating elements like big data, analytics, statistics, and the complex integration between all the elements to create a new form of interaction between people and machines.

Now, cognitive systems are commercially available and able to handle a growing list of tasks that were previously the preserve of humans.

Cognitive Computing in Financial Services

There has been an explosion of disruptive and digital transformative technology in the financial services industry over the last 3–4 years. The rapid growth of this technology sector has dramatically impacted incumbent firms which service the financial services industry. In 2016 alone, over $50 billion USD was invested in the fintech sector, covering 16 categories of technology and encompassing over 1126 companies. While the investment in pure fintech reached a peak in Q3 2016, an emerging segment within the fintech domain has begun to take on exponential growth, namely, cognitive computing.

Cognitive computing within the fintech sector is clearly viewed as the next focus area of the artificial intelligence wave of investments. While healthcare, advertising and sales, and security have been leading the focus of investments and deals in the last several years (CB Insights), in early 2016, the finance sector began to emerge as the next focus area for the application of artificial intelligence and cognitive computing.

The financial services industry, and specifically the wealth management segment, structures itself into three organizational areas where capabilities reside that enable the delivery of products and services to consumers: (1) front office, (2) middle office, and (3) back office. Each of these organizational areas can benefit from cognitive computing capabilities. These functional areas are overburdened with significant amounts of unstructured information, including a rapidly evolving regulatory burden.

Additionally, each of these organizational areas encompass roles that require specialized domain knowledge, such as financial planning, portfolio management, or KYC/AML (know your client/anti-money laundering), knowledge which is often embodied in specific individuals within a firm. Lastly, the middle and back office organizational areas are burdened with manual and labor-intensive processes that require human capital to achieve both financial and capacitive scales. The challenge that these characteristics present is where the benefits of cognitive computing provide the opportunities in the financial services industry.

Applying cognitive computing to applications in the financial services industry allows for the synthesis of vast amounts of unstructured economic, financial, and market data and can deliver valuable insight to support front, middle, or back office needs. This synthesis and insight can be guided based upon the objectives of the user of the insight. Cognitive computing can also supercharge the expertise of roles which require specialized knowledge, by allowing them to quickly expand their knowledge base. This is achieved by allowing the cognitive systems to learn what knowledge is required by the specific roles.

Examples of this learning might be heuristics that a human adviser follows to allocate assets in a portfolio based on experience; it might also be new government regulations. This learning narrows the information needed to empower the roles and provides insight and meaning to the information provided, allowing them to act faster with greater accuracy. At the same time, since cognitive systems can also learn, infer, think, and interact, experts within the firm can also "teach" the system through dialog-oriented natural language interfaces.

This teaching model allows the cognitive system to embody the expertise of the firm's most knowledgeable experts plus insights embedded in historical data, allowing for the democratization of the expertise throughout the enterprise, creating scale and consistency, and reducing risk. For example, historical data about customer online behavior suggested to one financial company more appropriate, individualized web landing pages for their different customers.

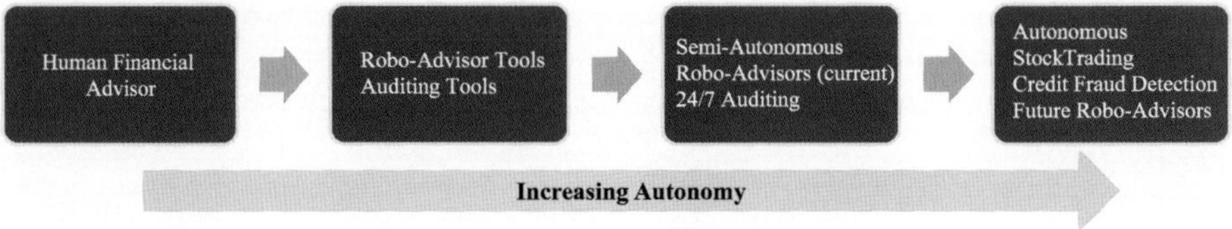

Fig. 2.1 A framework for financial tools, based on autonomy

Lastly, cognitively enabled business process automation (coined "smart robotics") allow for automation of operational tasks which would normally require human interaction. The notion of "smart robotics" moves the automation beyond binary decision-tree automation to full autonomous and self-learning operations. A conceptual framework for cognitive computing in finance might relate to the degree of autonomy of the cognitive system. As shown in Fig. 2.1, cognitive systems can range from assistants to financial workers, to completely autonomous systems.

Koch (2017) describes how a banking organization uses cognitive computing to predict overdrafts a week in advance with 94% accuracy and 4 weeks ahead of time with 87% accuracy. Chandarana et al. (2017) recently reported that banks "with the highest levels of digital execution saw front-office revenues per producer increase by as much as **eight times**, while those with the highest level of post-trade digitization posted **four times** more trades per middle- and back-office FTE than the bank with the weakest digital resources" (boldface ours).

Prior to the disruption caused by robo-advisors, these functions were considered part of the unique value proposition of the financial advisor and part of the advice delivery model. What the robo-advisors introduced was an automated and programmatic model to deliver these elements at scale. This has forced many advisors to redefine their value proposition to the consumer and to focus on the relationship model versus a transactional model. The future impact of cognitive computing will continue to enable this domain, as other dimensions of cognitive computing take hold within the advice delivery space.

The application of cognitive computing capabilities in financial services is on the verge of disrupting the industry. Areas that once relied upon human capital and human knowledge to deliver products and services are now looking at opportunities to apply cognitive computing to create scale (e.g., mass customization of products and services), efficiency, differentiation, and enhanced delivery of products and services.

Financial Advice and the Human Advisor

The financial services industry, and specifically wealth management, is anchored by various roles which deliver financial advice. These roles manifest themselves under numerous titles, including financial planner, investment advisor, investment manager, wealth advisor, etc. Generally, these roles are often bundled together and called an "advisor." The roles that generally fall under the "advisor" label are responsible for the delivery (and often the execution) of financial advice for consumers. To deliver the advice they provide, advisors may be certified or licensed by different regulatory agencies and maintain their certifications and licenses through examinations and regular continuing educational credits. It's through these certifications and/or licenses that advisors maintain a baseline level of technical knowledge that enables them to provide the products and services, including financial advice.

Knowledge Domains

However, beyond the baseline level of technical knowledge that the certifications and licenses require, advisors must be active knowledge seekers to inform their decisions, actions, and recommendations for their clients. The knowledge domains include, but are not limited to:

- Macro- and microeconomic conditions both in the USA and globally
- Regulatory requirements
- Tax implications and insight
- Investment market conditions and drivers
- Investment and financial instrument knowledge
- Information about products and services from various providers

In addition to the above domains, the advisor must also have knowledge and intimacy of the specific client's situation. This extends beyond basic individual information (i.e., age, amount of investable assets, etc.) into areas such as the client's appetite for risk, what specific goals and objectives they are looking to achieve, how they might prioritize and/or make trade-offs relative to achieving the goals and objectives, and other qualitative insights into the client's mindset (concerns, fears, perspectives, needs, wants, etc.). This knowledge also informs the decision process around suitability, which ensures the investment and advice provided is suitable to the client's individual situation.

Types of Advice Offered

The term "advice" is a broad definition that at a minimum encompasses investment advice and recommendations for specific investment allocations but is often extended beyond investments to include insurance, tax planning, estate planning, college planning, retirement planning, cash flow planning, and sometimes budgeting. The financial advice may also include helping the client get organized (financial and nonfinancially) and place emphasis on helping clients achieve specific goals and objectives.

Current Technologies Used

While the ability to deliver financial advice is predicated on an extensive knowledge base and intimacy of the client's situation, advisors also use various technologies to assist in the analysis of the financial data to produce components of the financial plan. For analyzing investments, the majority of these tools utilize a Monte Carlo method[1] of analyzing the data, which uses the underlying investment instruments, portfolios, and investment product data and then simulates various sources of uncertainty that may affect their value over time. These simulations are then run hundreds of times, and the tools map the distribution of the resulting value change through the different outcomes.

In addition, these tools also typically perform a client risk assessment, or risk tolerance analysis. This assessment asks a series of multiple choice questions in an attempt to assess the risk appetite of a client. The resulting "score" derived from this programmatic test is then mapped to specific investment allocations that map to the same score. For example, if a client receives a very low score (low risk, very conservative), that score would map to investment classes that are equally low risk, such as fixed income (bonds). The majority of tools that exist within the market utilize a combination of these models to aid in the delivery of the advice to the consumer.

Limitations and Opportunities

The delivery of advice to the consumer from a financial advisor is predicated upon three dimensions: (1) extensive market, regulatory, tax, and product knowledge, (2) extensive client knowledge and intimacy of the client's situation, and (3) the ability to utilize various financial planning tools to synthesize and analyze data. It's the combination of these three dimensions that enable the delivery of holistic financial advice to the client. Ultimately, the ability of the financial advisor to maximize their knowledge across the first two dimensions enables or inhibits their ability to derive insight from the analysis of the data and to deliver comprehensive and accurate financial advice. Robo-advisors can help tremendously in both tasks, but especially the first where they can be 100% current about numerous laws and regulations – and even client knowledge in a way that human advisors cannot, due to human limitations.

Advisors are inundated with external information of the markets, economy, regulatory changes, and product options. At the same time, as an advisor increases the number of clients they service, their ability to maintain the same level of knowledge and intimacy of every client decreases. The opportunity to utilize cognitive comput-

[1]The Monte Carlo approach employs computational algorithms using random sampling within probability distributions to obtain solutions for optimization and other problems.

ing to supercharge the advisor's knowledge and enable the advisor to synthesize the client's qualitative insights will allow not only for greater scale but increased accuracy and completeness of the advice provided.

Digitizing Advice

In late 2009 to early 2010, a small contingent of well-funded Silicon Valley technology startups were launched into the wealth management market which delivered "financial advice" to consumers using a technology-only approach, absent of any human advisor. These technologies quickly gained notoriety in the wealth management market and were coined "robo-advisors."

Current Digital Advice Offerings (Robo-advisors)

Robo-advisors are software programs, delivered via mobile or web experiences, that assist customers by delivering financial advice and customizing investment portfolios, typically without any human advisor input or engagement. While the depth and breadth of the advice provided was limited as compared to their human advisor counterparts, the robo-advisors exploded into the wealth management market and gained significant press and amassed assets from consumers quite quickly. Since the initial launches in late 2009 and early 2010, some 40+ robo-advisors have entered and disrupted the traditional investment and financial advice market. In 2015, assets under management (AUM) in the robo-advisors segment totaled $66 billion. This is forecast to rise to $225 billion by 2017 and over $1 trillion by 2021, according to Statistica (2017).

As this disruption occurred, it forced legacy brands and market incumbents to react to this new model of advice delivery. As a result, most major wealth management companies have adopted, or are in the process of adopting, some form of digital advice (robo-advisor) capabilities for their clientele. Some organizations, such as WealthFront, utilize digital advice exclusively, while others such as Vanguard, Fidelity, and UBS use digital advice in conjunction with human financial advisors.

Robo-advisors are transitioning from the management of customer financial portfolios to digital platforms. By doing so, they open opportunities for increased understanding of customers and improved risk management. They also offer increased appeal to millennials and digital natives who are more comfortable with computer-based access to services.

Types of Advice Delivered

While much is made of the intelligence and type of advice provided by robo-advisors, with few exceptions, all of the leading providers in the digital advice space utilize algorithmic solutions to portfolio management. They mostly support passive investors using ETF-based products and use buy-and-hold strategies. They offer dramatically lower costs to consumers, in the range of 0–20 basis points (bps) versus the typical 70–135 bps associated with human financial advisors. The robo-advisor market is now projected to reach $255 billion by 2020.

Techniques Used

Most digital advice (robo-advisor) technologies adjust balances between equities and bonds, international versus domestic stocks, and make their programmatic decisions based on various structured data inputs such as age, expected retirement date, amount to invest, the target amount needed, and aversion to risk. Many of the digital advice tools utilize models to show user-friendly projections of the investments being made and the likelihood of achieving their investment needs. They continually track markets, rebalance portfolios, and may feature tax loss harvesting. The majority of digital advice (robo-advisor) programs use a mix of:

- Modern portfolio theory (MPT), which maximizes expected returns based on tolerable investor risk.
- The Black-Litterman mathematical model for portfolio analysis developed by Goldman Sachs.
- The Fama/French three-factor asset pricing model. Eugene Fama and Kenneth French were professors at the University of Chicago Booth School of Business who incorporated risk, comparative performance of small and large firms, and book/market calculations to create their model.
- In-house expert opinion.

Market Impact

After a slow reaction to the disruption occurring in the market, the legacy and incumbent wealth management companies, such as Fidelity and BlackRock, have responded quickly to the pure play companies that deployed the first wave of robo-advisors. Along with many traditional financial institutions such as banks, the legacy brands and incumbents have now begun to adopt the digital advice technology to reach new client segments, service existing ones who desire a more passive and digital interaction, and have incorporated the technology into the enterprise to create scale and efficiencies for their human counterparts.

Cognitive computing is bringing new capabilities to the digital advice (robo-advisor) segment. While only a few of the current providers use machine learning, providers such as ForwardLane (powered by Watson) and Kavout now incorporate machine learning for automated research and natural language interfaces for their users. The new generation of cognitive computing-based digital advice platforms will be able to adapt to changes in market dynamics, learn more about investors, adapt to changing styles, and improve performance beyond basic MPT.

Applying Cognitive Computing to Advice

Traditional programs for portfolio management decision support leverage structured data, that is, data stored in relational databases. These data are stored in rows and columns but represent only a fraction of the data available to support investment decisions.

Structured and Unstructured Data Processing

New robo-advisors like Kavout and ForwardLane incorporate unstructured data such as text-based analyst reports and financial and economic news. These data are regarded as unstructured because they do not fit into typical database structures, e.g., tables and rows, and are usually nominal, not categorical, data. Kavout and similar advisors ingest unstructured data from many sources and build overlapping neural network models using that data. They then use those models to improve portfolio analysis and decision-making.

Natural Language Processing and Speech Synthesis

Natural language processing is another key feature of the leading robo-advisors. This enables users to interact with the advisers using spoken language, making them much easier to use. Natural language is digitized, converted to text, tokenized, semantically interpreted, and used to query existing knowledge structures within these advisors.

Some of these advisors also have the ability to communicate with their users via natural language. This feature is speech synthesis – the artificial production of human speech. Again, just like natural language understanding, speech synthesis reduces the communication barrier between human users and advisors.

Sentiment Analysis, Making Inferences, and Recommendations

Sentiment analysis brings expressed sentiment in unstructured data into play in portfolio management. Unstructured data such as analyst reports can be analyzed by the expressed sentiment and the strength of the sentiment regarding specific stocks or markets or companies. This has become a powerful tool enabling the addition of new variables to the process of portfolio management.

Ultimately, the strength of these cognitive robo-advisors in wealth and asset management is that they can make inferences, evaluate the trade-offs between different decisions, and offer recommendations. It is this reasoning ability, coupled with the ability to draw upon vast amounts of data, that gives robo-advisors their immensely powerful capabilities and future promise.

Future Impacts of Cognitive Computing in Financial Advice

Today's use of cognitive computing in the delivery of advice is in the early stages of adoption. A small number of platforms exist which are applying a limited number of capabilities that fall under the artificial intelligence and cognitive computing definition. The most common applications today include sentiment analysis, predictive and prescriptive analysis, and natural language processing. As the depth and breadth of adoption of cognitive capabilities increase, the ability to impact all of the related parties within the value chain of financial advice delivery will accelerate and dramatically increase.

Platforms will expand on their utilization of NLP (natural language processing) to understand what the clients are concerned with, what their priorities are, and what their goals and objectives are. They will be able to know the sentiment of the consumer and where trade-offs and recommendations can be made, and the systems will offer those in context to the learning that the system has garnered from the advisor themselves. At the same time, the cognitive systems will expand further into the knowledge domains, enabling advisors to break through their cognitive limits and have access to synthesized and insightful data that will enrich the advisor's relationship with the consumer.

How Will Advice Be Generated?

The future of cognitive computing for the consumer is one that will continue to advance as well and quite rapidly. The advances in NLP (natural language processing) will continue to allow natural language interactions through cognitively enabled chat bots that move away from preset, programmatic answers to contextual responses based upon the consumer's input. There will be a transition for the consumer from a transactional-based model with current digital advice platforms (robo-advisors) to a relationship-based model. The systems will be expanded in their ability to learn about the client's unique situations and combine that knowledge with information external to the client, such as the economy, product insight, or investment market conditions. This convergence of data, specifically unstructured data, will enable the next generation of robo-advisors to deliver even better digital advice and expand in use as customers become more used to them.

The technologies will be enabled to interact with a consumer in ways similar to that of their human counterparts, asking questions that allow adjustments in recommendations and the ability to make inferences to help the consumer think about options. This iterative model, similar to a face-to-face meeting with an advisor, will allow the cognitive-based systems to adjust recommendations and the advice they deliver with full consideration of risk, suitability, external market conditions, and the unique client situation.

Impact on the Customer

Ultimately, as the advances in cognitive computing continue, there will be greater positive impacts to the consumer. Consumers who desire a more digital engagement augmented by financial advisors will be able to receive comprehensive and accurate advice, at an economic value that the consumer demands. At the same time, consumers who wish to engage in a face-to-face advisory relationship will have access to human advisors who are more knowledgeable and can provide greater insight and advice.

References

Chandarana, D., Faridi, D., & Schulz, C. (2017). How cognitive technologies are transforming capital markets. *McKinsey & Company*. Retrieved from https://www.mckinsey.com/industries/financial-services/our-insights/cognitive-technologies-in-capital-markets

IDC. (2016). Worldwide spending on cognitive systems forecast to soar to more than $31 billion in 2019. *IDC*. Retrieved from https://www.businesswire.com/news/home/20160308005344/en/Worldwide-Spending-Cognitive-Systems-Forecast-Soar-31

Koch, R. (2017). Cognitive computing: Teaching computers to learn. *Strategic Finance, 98*(10), 62.

Reynolds, H. (2016). What can self-driving cars teach about cognitive computing? *KM World, 25*(10), 40.

Roe, C. (2014). *A brief history of cognitive computing*. [online] Dataversity. Retrieved from http://www.dataversity.net/brief-history-cognitive-computing

Shanks, R., Sinha, S., & Thomas, R. (2015). Managers and machines, unite! Three things managers must do to make the most of cognitive computing. *Accenture Strategy*, 1–7.

Statista. (2017). FinTech. *Statista*. Retrieved from https://www.statista.com/outlook/337/100/robo-advisors/worldwide#contentlist

Williams, S. (2016). Outthink cognitive hype: Creating a business-driven cognitive strategy. *Business Intelligence Journal, 21*, 28–36.

Chapter 3
Living or Dying in the Mashup of American Financial Services: Literate Does Not Mean Competent

Elven Riley and Mark Schild

Abstract Clients are aging, passive investing is gaining favor, and client objectives are growing in detail and diversity. At the same time, advisors themselves are reaching retirement age with no clear succession plan in place. Technology disruption waves are transforming accounting, insurance, and estate planning, but it is the investment advisor who needs to adapt now or face extinction. We examine the risks facing the entire industry by focusing on the small independent advisory firm's role and place in the ecosystem. We suggest that technology will accelerate the growth of the financial advising industry at the same time that the wealth accumulated by older generations migrates slowly to younger generations. We underscore the natural advantage of the small independent investment advisor as the trusted partner to a client's life-cycle decision-making. We suggest that the small independent advisor can only survive and thrive through reinvention. The preferred trust relationship will remain between human client and human advisor but will be heavily machine augmented with analytics and big data, delivered via the cloud.

Keywords Robo-advisor · Generational wealth management · Augmented human teams · Machine learning · EGADIM · CFP · Life-cycle advising · RIA

Introduction

There have always been threats to the survival of the financial service industry, prognosticators forecasting its demise, and scandals tarnishing its reputation. The current threat is large firm dominance of the registered financial advisory business using technology, eliminating the small firm niche entirely. The largest firms are investing in unproven artificial intelligence and machine learning delivered as mobile phone applications (robo-advisors) reaching millions of clients to capture assets under management (AUM). Millennials, portrayed as abandoning traditional

E. Riley (✉) · M. Schild
Seton Hall University, South Orange, NJ, USA
e-mail: elven.riley@shu.edu; mark.schild@shu.edu

M. Anandarajan, T. D. Harrison (eds.), *Aligning Business Strategies and Analytics*, Advances in Analytics and Data Science 1,
https://doi.org/10.1007/978-3-319-93299-6_3

advising just as generational wealth transfer begins, are responding but have little wealth currently. New market products, such as exchange traded funds, are touted as the replacement for portfolio asset management, and political rhetoric drives overreaching regulatory reporting, stifling the entire financial services industry by adding uncertainty to the horizon.

We analyze an industry adapting to client-shifting needs and adopting the latest tools available. The scope of our topic covers the landscape of financial services, but we will anchor our analysis to specifics in the registered investment advisory (RIA) community. We highlight the opportunities in the RIA industry in contrast to recent doomsday headlines and believe the changing demographics of both clients and advisors are a distraction, not a force. We see startup financial technology companies introducing robo advising applications weekly, devoid of traditional product fees and based on unsustainable business models and destined to collapse. Finally, caught in the vice grip of both regulatory and competitive change, advisors must again be proactive and adapt or become irrelevant.

Beginning with the investment advisor (IA), we highlight the urgent need to address the aging population of advisors, recruit younger advisors, and build teams of experts to address clients' varied needs. We also address the need for the IA to become technologically competent in order to stay ahead of their client's technology literacy. Next, we outline how advising, anchored in a life-cycle timeline of financial decision-making, supports each client's financial literacy. We will show that the timeline does not change although the client's place on the timeline shifts with age and personal conditions. Finally, we will address the fact that the IA and the client must be better aligned. While clients are looking at inexpensive financial service alternatives to solve their needs, advisors remain committed to traditional expensive fee-based investment services. The advisor needs to become part of a team, offering holistic planning advice and utilizing cutting-edge technology. We will show that a team with technology will always beat a team without technology. Equally important, we will show that a team with technology will always beat technology without a team.

Hypothesis

The above observations motivate the following three hypotheses:

1. IA technology literacy is required and mastery is preferred to enable a sustainable business model.
2. Client financial literacy is a required outcome of the advising process to capture the increasing wealth accumulation throughout the client's life cycle.
3. A machine-learning augmentation applied to a multi-faceted IA team is better suited to meet the industry's and the client's ultimate requirements.

Let's Meet the Advisor

The Growing Industry

The US Bureau of Labor Statistics (BLS) defines personal financial advisors as those who provide advice on investments, insurance, mortgages, college savings, estate planning, taxes, and retirement to help individuals manage their finances (Bureau of Labor Statistics, 2018). Interestingly, the Certified Financial Planner (CFP) Board describes itself in a very similar fashion. "From budgeting, to planning for retirement, to saving for education, to managing your taxes and your insurance coverage, 'finances' doesn't mean just one thing for most Americans — and 'financial planning' means much more than just investing. Bringing all the pieces of your financial life together is a challenging task" (CFP.NET, 2018). More on the CFP and their role will be discussed later.

In reviewing industry growth, BLS estimates that there were 271,900 financial advisors employed in 2016 and the 10-year growth, estimated at 14% (2016–2026), is double the 7% average growth rate for all occupations. This growth is not surprising. As BLS indicates, "the primary driver of employment growth will be the aging population. As large numbers of Baby Boomers approach retirement, more are likely to seek planning advice from personal financial advisors. Also, longer lifespans will lead to longer retirement periods, further increasing demand for financial planning services." That's the good news (BLS, 2018).

The Aging of the Industry

A more sobering fact, a 2016 report by Cerulli Associates states that 51% of the country's financial advising professionals are 55 and over. As for younger advisors, the same report states only 26% are under 44, and Ernst & Young (EY) indicates fewer than 5% are under 30. In the same report, EY also indicates that younger clients desire younger financial advisors (Fava, Boersema, & Hamaloglu, 2016). Equally troubling is that Kevin Keller, the CFP Board's chief executive, stated that "if we don't take action, we're going to find ourselves with a significant shortage of financial advisers" (Benjamin, 2017). Knowing there is a talent literacy shortage in the RIA business, Scott Hanson, of Parthenon Capital Partners, inquires at Sacramento States' CFP training program and found that only 3 of 20 graduates are headed into planning, but the rest see no career path (Southall, 2017). Clearly, students are not being exposed to opportunities in the financial sector. Despite the aging of the industry, the 2016 Evolution Revolution profile of the IA profession shows the industry is still enormous and growing. In 2016 there were over 11,000 SEC registered advisors (3.5% over 2015) with over $66 trillion in assets (Barr & Gebauer, 2016). However, this growth is deceiving as over 72% of the assets are managed by the largest 3% of the advisory firms. The IAA/NRS report also shows that the industry is predominantly small businesses, with the majority (87.8%) employing 50 or fewer employee. The larger firms are already building teams to address multiple client-planning issues, as well as advisor-succession planning. As assets get more concentrated within the larger firms and the overall industry professional ages, it will be more difficult for smaller RIA firms to adequately recruit (Armstrong, 2017).

In order for the advisory firms' advice to be effective, client and advisor need to be in sync and effectively communicating. The spring 2017 Merrill Edge report highlighted that people feel more comfortable with advisors that are of similar age or at least who treat them as individuals, rather than as a stereotypical demographic. For instance, Millennials, as a group, are likely to have several jobs over their lifetime, may decide to rent versus own, and expect to get answers with the click of a button. However, many in that generation, born in the early 1980s, are burdened with a staggering amount of student debt and have had difficulty finding a job or getting raises during and after the Great Recession (Levine, 2017). The same financial strategies that worked for previous generations may work for younger generations, but not now and possibly years in the future. We all move along the same timeline, however, where we are on that line shifts. The desire to work with someone of a similar ilk is unsurprising and is simply another reason that shifting from individual advisors to teams containing advisors with varying expertise and demographics is one of our recommendations (Tibergien & Dellarocca, 2017).

To summarize some of what we have discussed so far, there is still a need – in fact, a growing need – for financial advice. The age has changed since people are living longer, working longer, and potentially starting their adult life in debt. As the wealth transfer continues, over $30 trillion will pass to the next generation (Accenture, 2015). People need advice on how to manage that money. Michael Collins (2012) reemphasizes what J.A. Haslem (2008) asserts: "financial advisors can help clients overcome feelings of insecurity, help validate clients' past decisions, and serve as a neutral party in spousal disagreements." More good news is that studies commissioned by the CFP Board and conducted by the Aite Group indicate CFP professional practices include a 53% higher share of both high net worth and ultrahigh net worth clients. The reason given is due to "…disruption in wealth management in general and a pressure on advisors to deliver more value for the price they charge…folks that deliver a planning-centric model to their clients are able to provide that value proposition that [clients] are looking for" (Pirker & Schmitt, n.d.).

Despite job opportunities in financial advising, Millennials with or without 4-year degrees are not rushing into the industry. There is little media attention to the financial advisory industry other than how boring selling life insurance can be. Halah Touryalai (2012) states there are barriers to entry compared to the 1990s when Wall Street brokerage firms, like Merrill, UBS, and Morgan Stanley, served as significant training grounds for young financial advisors. Additionally, Touryalai points out that older advisors started out in an age of cold-calling, which college grads are not attracted to and, frankly, neither are the clients. A study from

Kenan-Flagler Business School, University of North Carolina (Brudner), identifies over 80% of decision-makers will not buy from a cold call (Brudner, 2016). The drive for automation of administrative tasks and industry consolidation masks the labor shortage at the moment. A lack of entry-level employees is potentially a significant problem and could choke growth and profit. Looking at the industry's demand for a pipeline of young candidates spotlights the opportunity for focused educational programs. The demand for financial advisors is expected to grow by 41% in the next few years according to *Money* magazine and Payscale.com. But how can companies attract young people to this industry? Companies such as Merrill Lynch, Fidelity, and TD Ameritrade are all offering grants, scholarships, and jobs to new graduates pursuing the certified financial planning certification, and the number of CFP educational programs continues to grow. Seton Hall University Stillman School of Business, a private 4-year university, has begun the approval process for a CFP program. However, if the slow adoption of CFA-certified programs and the CPA programs that come before is any indication of pace, the industry may need to look for alternative employee pipelines.

Let's Meet the Client

Which Age Groups Have Investment Questions?

The Silent Generation, Baby Boomers, Generation X, and Millennials all play a role in our economy, affecting what we eat, what we watch, where we travel, and yes, how we manage our finances. Public media would have you believe that the Millennials (people born after 1980) have taken over and won the battle for economic dominance. We believe there is still time. Studies from EY, PwC, Merrill, the Federal Reserve Board, and Department of Labor all indicate that Millennials are deferring decisions about home and car purchases, as well as marriage. The biggest reason given is the Great Recession and its effects on Millennials.

As Fig. 3.1 shows, the numbers of non-Millennials decline quickly over the next decade, and although that may be true, by 2028, there are still expected to be over 140 million non-Millennials eating, traveling, watching television, and managing their finances. Nielsen states while Millennials may have surpassed Boomers as the largest living generation, Boomers still control about 70% of all disposable income in the United States. But as Tony Stich (2017) points out, do not forget Generation X, as they are on the front-end of the wealth transfer and are currently in a "retirement saving mindset." Additionally, Stich states that, "…often after years of neglect... [Gen-X] will expect an experience on par with robo-advisors, but more importantly, one that provides personalized advice and the knowledge to back it up." In fact, according to data from Pew Research, as of 2012, 36% of young adults (aged 18 31) were still living with their parents (possibly those Gen-Xers) and only 25% were married (Fry, 2016). Additionally, the Bureau of Labor Statistics data

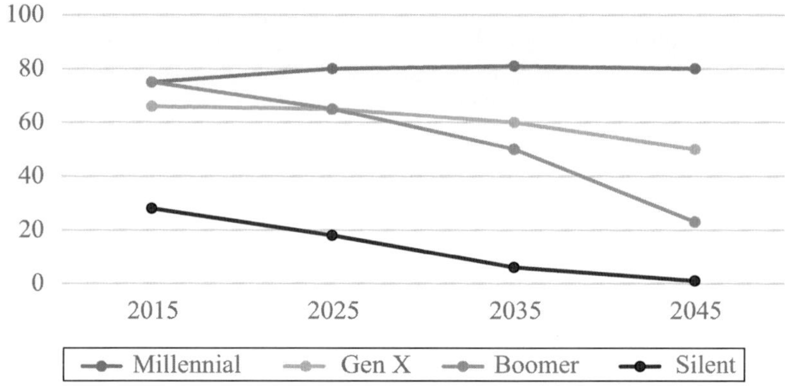

Fig. 3.1 Millennials overtake boomers (population by generation)

indicate an unemployment rate of over 14% for young people (ages 16–24) in 2014. The 18–31 age group could not afford to rent apartments, furnish homes, and certainly not invest in markets. Therefore, it is no wonder that a 2016 study by PwC and Bankrate.com states that only one in three Millennials have money in the stock market (Cornfield, 2016). Even older Millennials, who have more capital available, state in surveys that there are greater demands on their money, as they want to start families, buy homes, travel, etc. Also, per the spring 2017 Merrill Edge report, Millennials, as a group, are not necessarily looking at finances in the traditional sense. Merrill finds over 60% are looking to save to live their desired lifestyle, rather than a complete exit from the workforce. In fact, in a survey of over 1000 affluent Millennials across the United States, the survey indicated Millennials' willingness to work for the rest of their lives to achieve freedom and flexibility (Shekhtman, 2016). However, Millennials also admit to being unprepared for life's "what-ifs," like divorce, children, or outliving their assets (Berridge, 2014). The good news is that over 60% of Millennials in the survey save 5% of their income, up from only 42% saving in 2015. Even better news, according to EY, the oldest Baby Boomers have begun wealth transfer to their children, and by the end of this decade, Generation X and Y investors will accumulate close to US $46 trillion in assets. So there is time to build those relationships with the next generation of investors (Jain, Hyde, & Lyman, 2013).

As Fig. 3.2 shows, the money is shifting slowly. Millennials may be making all the noise, but Boomers still have much of the money, at least the investable money (Baby Boomers, 2016). Millennials are influencers, and the robo-advisors are growing AUM simply based on Millennials' desire to do things themselves and to spend as little as possible in the process (Harding & Allison, 2016). As Fig. 3.3 shows, in recent government statistics, people 65 and older, with little savings and surviving on Social Security, are forced to put off retirement and stay in the workforce longer.

Whether advising Boomer, Generation X, or Millennials, advisors should ignore the labels and focus on the financial advice each generation requires to

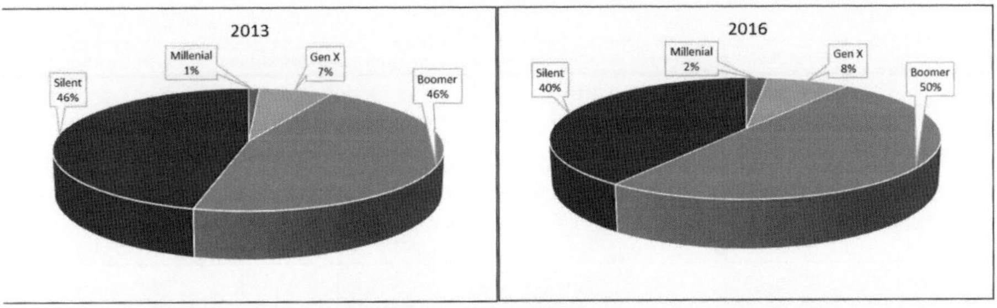

Fig. 3.2 Silent generations' wealth transferring to boomers

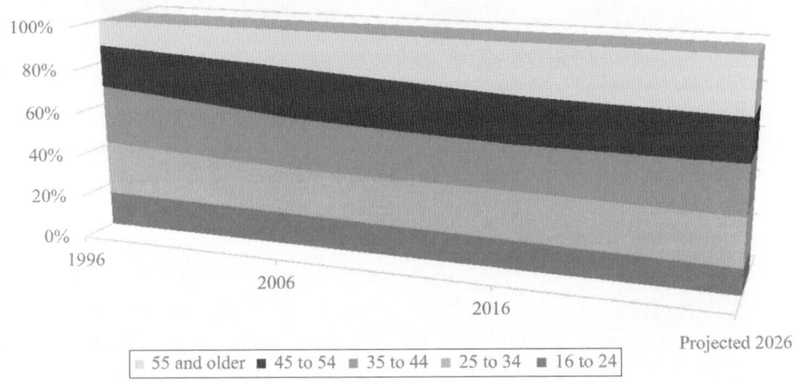

Fig. 3.3 Percentage of labor by age

define and redefine their goals. As Cam Marston stated in his Generational Insights, Millennials are expected to control more than $11 trillion by 2030 and inherit over $340 billion by 2040. Although we do not focus on it in this paper, the Bank of Montreal's Wealth Institute survey finds women control about 51%, or $14 trillion, of American personal wealth. Wherever and whoever has the assets, they need financial advice. Although each is at a different point in their life, everyone has the same life-cycle questions.

The Financial Life Cycle of a Client

How should they pay for education (learn)? How should they save and invest their money (earn)? Do they have enough to retire (burn)? Can they leave anything to the next generation (legacy)? (Fig. 3.4).

Whether advising Boomer, Millennial, or Generation X, there is no difference. They all have a wide variance in personal goals that change as they age. The world is not necessarily more complicated now than in the past, but the speed with which

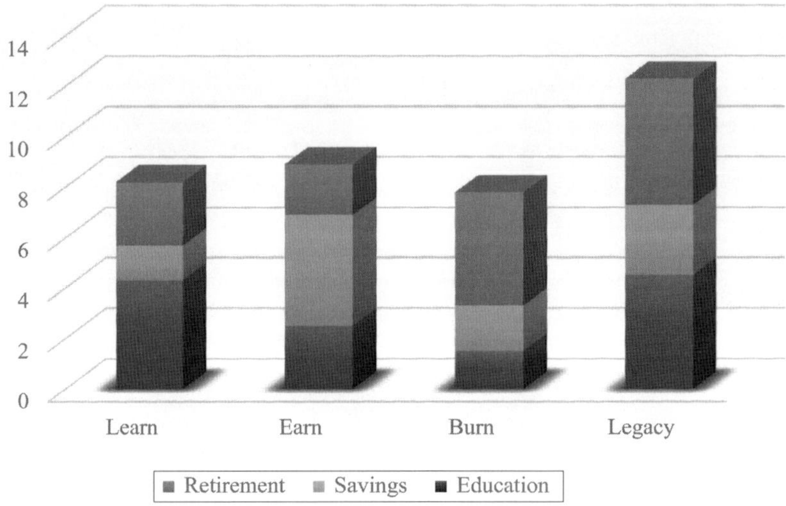

Fig. 3.4 Life-cycle progression

everything moves and the immense amount of money in motion make decisions more impactful and potentially more damaging. Increasing risk without increasing return neutralizes today's static strategies. The client must be a part of the process and that takes effort. The earlier parents include children in decision-making, the better. As Deena Katz, co-chairman of Evensky & Katz/Foldes Financial, states, "our country is woefully, financially illiterate…It hasn't helped this generation that their parents didn't teach money management and often tried to hide their own financial difficulties" (Mellan & Christie, 2017). While parents are rethinking retirement and legacy planning, Millennials need to know the details of their education funding and perhaps plan for debt servicing strategies.

The age of the client helps prioritize the goals and possible objectives. These life-cycle goals are named and are well known to the advising team members. Financial planner, IA, accountant, estate planning attorney, and insurance agent are all required for solutions to meet the stated goal. The services are aligned by financial industry provider, very often with multiple certifications and license designations.

But if financial literacy is not the responsibility of the educational system and elders are not coaching the next generation, then where do people of all ages go for advice? As can be seen from Table 3.1, life-cycle planning covers many areas throughout a person's life. Add to the complexity that each client is unique, so the plan should also be unique. In the chart above, all of the goals could require traditional investments, risk management (insurance), tax planning, and legal advice. But if the client is illiterate and the advisor is unaware, how does the financial decision get made? Know Your Customer (KYC) began with regulatory responsibility for anti-money laundering (AML) but now leads to a deeper KYC, developing and changing planning strategies. We see technology as a huge help in data gathering, aggregation, and analysis. Consumers, regardless of demographic, will continue to demand information faster, cheaper, and with greater transparency.

Table 3.1 A sample of client goals, services, and product types and the team members

Goals	Services	Vehicles	Team
Education (self)	529, Series EE/I, prepaid tuition	Mutual funds, ETFs, strips, whole life	Financial CFP, insurance CLU and CPIA
Marriage, estate planning	Wills, trusts, living wills	Insurance – life, health, property and casualty	Attorney CELA, insurance CLU and CPIA, accounting CPA, financial CFP
Children (planned and young) care and education	UGMA, UTMA, 529, prepaid tuition, trusts, Coverdell	Strips, whole life, mutual funds	Insurance CLU and CPIA, financial CFP, accounting CPA, attorney CELA
Income augmentation	Investment portfolio, tax planning, risk assessment	Stocks, mutual funds, fixed income, alternative investments	Financial CFA and CFP, accounting CPA
Retirement (self and spouse)	Health plans, inheritance plan	Annuities, IRA, 401 K, SEP, REIT	Financial CFP, accounting CPA, attorney CELA, insurance CLU and CPIA

A Computer Wants My Job

Over the past 20 years, technology tools have infiltrated every aspect of business, and the IA is no exception. As the large wirehouses migrated to a standard PC-based desktop, the tools of the trade expanded from voice telephony to include email, spreadsheets, and charting. The independent IA moved more slowly, replacing paper files and hours on the phone but not necessarily the dinner hour house calls or the rounds of client golf (Fig. 3.5).

CRM

Today, successful IAs growing their client base rely on customer relationship management (CRM) systems to assist them in remembering each client's details and managing their time more effectively. Conversion of a client's interest in a new service to a productive revenue stream is the combination of sales skills and attention to details. With a small number of clients, many IAs memorize every client fact from birthdays and children's names to property addresses. With a larger number of clients per IA, only the rare photographic memory could survive, while the remainder decrease their perceived value to the client. The now ubiquitous CRM solutions provide all of the client data, delivered via the relationship style of each IA. There are many examples of CRM systems from Salesforce, Wealthbox, Redtail, Envestnet, Junxure, Zoho, and HubSpot, to name a handful. The costs of these solutions have plummeted, and only the initial conversion of disorganized digital and paper records

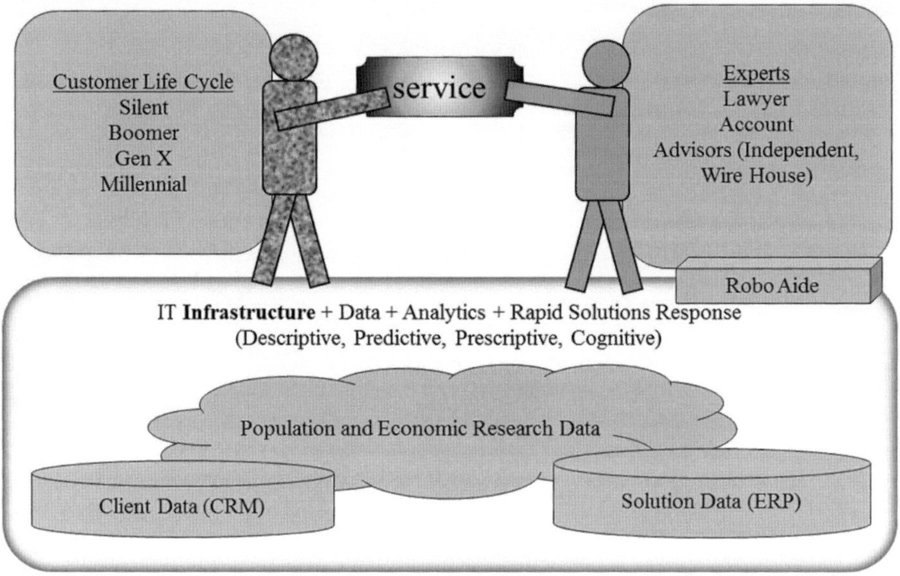

Fig. 3.5 Human-to-human service delivery requires infrastructure: more than the combination of ERP and CRM systems

to a consistent form remains a challenge. The benefit to the IA is that firm-wide, client information is uniform and current from service to service eliminating constant administrative overhead costs and embarrassing small errors (misspelled middle name on a new service on-boarding contract). The consistent and up-to-date information is also essential when regulators (SEC/FINRA) show up.

The training and integration of new IAs take effort and tax the senior advisor's time. This remains true today as older advisors guide younger advisors into the profession. Neither is revenue productive while the training process is under way. Time-to-first revenue is reduced by a CRM providing the service history and current details of long-time clients. Retention of new IAs is increased as their activity becomes more self-directed and monetarily rewarding.

ERP

As CRMs automated the capture of client information, the enterprise resource planning (ERP) systems not only housed the information but also automated and standardized the operational work processes within the organization. The large wirehouses were the automation leaders, and as the technology price point became more accessible, the independent IA joined in the automation wave. Products from Netsuite (Oracle), IBM, SAGE, SAP, and Microsoft offer many choices and prices. Custodians of client assets, like Pershing (Bank of NY), Fidelity, Schwab, and others, offer automation of the client statements by automating the service transactions. Moving from dozens of papers with signatures to screens verifying information entered has helped turn a necessary but tedious task into an automated one. Moving from order to execution with no human intervention, named straight-through processing (STP), has displaced many operational staff positions while simultaneously increasing transaction volumes.

Big Data

Today, the industry is building on these foundational efforts. The engineered large data capacity and multi-processor computer speeds are a commodity. The costs have continued to drop and the next wave of disruption has begun. Data from both the CRM and ERP systems are being enriched with other internal operational systems such as the legal, regulatory, and HR information unique to each IA. Again, familiar names from IBM to SAP and Oracle are providing these services. Cross-selling was the intention in the large wirehouses in an effort to catch up with the independent IA multi-asset class solutions. As more and more data were integrated, including external sources, the discovery of new insights into existing lines of business propelled a new initiative. These "big data" efforts now integrate massive amounts of information including external sources as diverse as social media, enabling accessibility never thought possible. The IA today can provide the client with analysis more detailed and nuanced than produced by the best market researchers 10 years ago (Bertolucci, 2013).

Machine Learning

The future is in harnessing machine learning, still an infant with great promise. The robo-advisor released into production is extremely narrow and limited. To complete even simple transactions often requires significant manual effort and unusual patience by the client. Where machine learning can be impactful is in the development of formal decision trees, a technique for reverse engineering how decisions have been made historically and predicting what the next decision step should be to successfully complete a transaction. Coupled with big data, the next step could be to suggest the IA begin final account review. The machine learning, reasoning, and logic are capable of adapting with more history, while the static prescriptive logic found in an ERP must be reprogrammed when conditions change. The proactive suggestions may be most effective with newly recruited advisors, helping them gain confidence at this early stage. In fact, as Benjamin L. Britton asserts, there is "...a greater demand for analytical abilities in junior recruits as the cognitive understanding of what data represents is a weakness of artificial intelligence (AI) thus strengthening augmentation between employees and technology" (Britton & Atkinson, 2017).

The Commodity–Priced Robot

We note the large investments being made in automation and especially in robo advising. We suggest the greatest ROI from automation is found in the quality and depth of the client profile data, not necessarily the application tool used to collect the client data. A great example of building a tool at great expense only to give it away is the Aladdin risk analysis and asset management system built by the world's largest asset manager, BlackRock. CEO Larry Fink confirms that BlackRock began building Aladdin in 1988, the year the firm was founded, and that he believes within 5 years, more of the company's revenue (not necessarily profit) will come from software and technology than from traditional money management. Fink has said, "We're using technology to be more connected with our clients, to help with financial literacy, to help all our clients navigate all our clients' money" (Zulz, 2017).

Market shifts and media hype once drove clients (and fees) to the professional portfolio asset manager. However, as Wharton Professor Jeremy Siegel points out, robots are just better at analyzing data and picking stocks. Professor Siegel states, "A manager must produce 10 years of market-beating performance to make a convincing case for skill over luck" (Wharton Executive Education, 2018). The combination of computers, with a digitized history of human decisions, and well-defined game rules has led robots to beating individual humans at chess, Go, and Hold'em Poker. White collar automation will displace humans just like the original calculating job displacement gave "computers" their name. Maybe Dell Corporation will begin offering a line of "equity analysts." There is no question Alexa from Amazon and Google Home will be offering basic financial services one of these days. Meanwhile, human advisor financial advice involves more skills and ambiguous game rules, which are well beyond AI capability. In addition, as Darren Tedesco (2015) points out, "many advisers love learning about and selecting individual investments; some even feel that the investment selection process is the core differentiator they bring to the table. Plus investment selection is what they grew up doing and one of the things that attracted them to the business model." The problem is that the modern-day advisor needs to be an asset gatherer, not an asset manager. Tedesco continues to point out, "…there's nothing with that [investment selection], at least not until it becomes a competitive disadvantage. The best 'pure' robo-advisors today literally spend 5 cents per client per year managing and maintaining their portfolios (including trading costs and so on)" (Tedesco, 2015). Besides the cost factor, the active managers' performance is simply not competitive.

A Computer Wants to Know Me

Over the last 20 years, technology has been entwining itself in the workings of business, and over the same 20 years, technology has also been symbiotically establishing itself in how we communicate. The paradoxical reality that Millennials rarely speak audibly to another human using their mobile device underscores the communication change. But, now even the Boomer clients have come to expect, sometimes with apprehension, to be recognized and understood by these robo-advisors. There is a decreasing number of phone call solicitations and snail mail sales offers from IAs. The twenty-first century's communication infrastructure enables an enriched conversation between IA and client. A concept that once required a day to explain to a client is now a simple interactive graph, allowing the client to change the input and watch the projected cash flows adjust. What used to entail reams of paper for each presentation can now be uploaded to an iPad and reviewed (and edited) in real time with a stylus.

Arming the client with the same data and analytical toolbox that the IA is using will generate more informed client questions and more verifiable answers and ultimately establish trust in the IA, who can bring context to the clients' life goals. Accepting current trends, one of the world's largest brokerage firms is about to deploy a robo advising strategy. In a May 2017 Bloomberg article, Morgan Stanley indicated it was set to augment its 16,000 financial advisors with machine-learning algorithms (Son, 2017). As the article points out, the best hope human advisors have against robots is to harness the same technologies that threaten their disruption: algorithms combined with big data and machine learning. In fact, as Michael Kitces, partner and director of research for Pinnacle Advisory Group, said, in terms of advisors adopting B-to-B "robo" solutions, "advisors have had rebalancing and trading software for a decade" (Kitces, 2016). Robo-advisors are gaining assets, but which generation's assets and are the assets new or are the assets being relocated from a human IA?

When viewed from the client's perspective, the IA relationship becomes all about the client. The technology decision is the client's. The decision to share personal information is the client's. The prioritization of life-cycle events, marriage or children, is the client's.

My Technology

The communication channel between the client and the IA is now heavily digitized with the client providing their own device and the IA extending their technology to the client. Flexibility is traded against security with the IA assuming most of the responsibility for security breaches. In fact, one of the most common phone calls to clients now is the required confirmation that the client's email requesting funds is actually from the client, not a hacked fraud. Given the dubious state of a client's technology, the prudent decision is to minimize the reliance on the client's technology. This is accomplished by either limiting communication to email-type messages or providing all interactive services from a cloud-based application. The cloud paradigm includes using hosting vendors to provide an integrated communication solution fully encrypted, secure, reliable, and available on an infrastructure the IA can access and pay for. Hosting services from major suppliers like Google,

Amazon, and Microsoft provide Software-as-a-Service (SaaS) solutions that many advising firms lack the technology staff to build. One of the largest value-added service custodians of assets (Pershing, Fidelity, Schwab, etc.) provide is the safekeeping of assets including data aggregation reporting and real-time quotes and trading interfaces. Between SaaS and the value-added service providers, the large firm with a large technology staff is challengeable by the small firm with the small technology staff.

My Personal Information

There is an overdue reckoning in the financial industry and social media about which personal data the service provider can use and which data cannot be used or sold. While we await the regulatory clarification, the focus is on asking for permission to broadly capture both static and dynamic personal client information. Once permission is granted, the data flow into "tagged" fields. The purpose of a tagged field is that the tag defines the data; for example, an "LN" tag signifies that the field contains the client's last name. Your digital identity is established using a number of tags, some which can easily be found in the public space and some that are kept private. Knowing the client and knowing all of the client's asset holdings are required to provide quality financial advising. Much of the CRM and ERP applications effort have been focused on collecting and maintaining these two information sets, identifying the client and knowing the assets. Both client and advisor continue to be increasingly dependent on technology, while also remaining incredibly nervous utilizing technology (Fiegerman, 2017).

Two new developments are now underway. The first is the sourcing of external information into the client's profile. This source information can be as simplistic as the average income statistic reported to the federal government for the client's residential zip code or as complicated as the client's job history extracted from a LinkedIn profile. The second development is the encoding of behavioral decision history. This history includes when decisions were made, as well as the choice made from a selection of options.

The use of these two new data enrichment sources enables the use of simple machine learning – heuristics – which can identify patterns and associations. Much like the algorithm music recommendation of "if you liked this song then you might like this other song," these suggestions have higher predictive hit rate with an increased variety of data sources and the increased historical decisions to draw on. The IA bots are capturing AUM and new clients, but their significant capture is the personal behavior data of each captured client. The IA bots are building a decision history unique to each client profile while simultaneously aggregated for association patterns just like music pick lists.

My Digital Life Cycle

Beyond the detailed personal information is the context provided by knowing the client's life-cycle stage. As we have discussed previously, clients need (and want) a new advising style focused on client life goals. Some call it financial life coaching (think Tony Robbins). Advising is now about everything from birth to death. Associating the millions of individual situations and navigating various client life decisions helps bring the power of statistical actuarial analysis to assist in answering even the simplest of today's client question: "I have never invested before, what's an ETF?" The engineering of the models to properly associate all of the relevant data tags and model the historical decisions with weighted positive outcomes is devoid of the human touch. Associating today's client decision with the life-cycle stage of the client adds a weighting of priorities more appropriate for the typical focus of the client. By doing this, AIs can avoid a client exclaiming, "Why do you continue to try to sell me an assisted living contract when I am focused on paying for an engagement ring?" The refinement of life-cycle information is not easily deduced from captured data streams of web browsing. Priorities and life goals are fuzzy-ordered, not hard-sequenced, and change as the client's life changes. Gut feel and intuition are still primarily human advisor skills.

Brain research into communication of goals (Lustig, 2017), using magnetic resonance imaging data, indicates much work must be done to further our understanding of how humans finalize decisions before automation can be applied. Verbal decision data do not always accurately track with the actual behavior recorded. The area of AI displacing human IAs remains the subject of movies and fiction, for now. The simple age classification, based on birthday, is a start, but after that, the ability to classify a client's life-cycle stage requires human judgment and interaction, with the risk of still missing the client's financial goals.

However, once the client life cycle is identified, as Table 3.1 illustrates, the resources that can be promoted, either as services or to support product transactions, are numerous. Identifying a client's need for tax planning, wills, college funding, or rollovers, just to name a few advisor services, remains as important an enabling function as ever. The identification of the subject matter experts required to support a client through the next decision task can be mapped as easily as a supply chain analysis for manufacturing and a sale of tomato sauce. An important contribution that automation can provide is identifying the perfect timing for client-advisor introductions, based on clients' immediate needs and wants.

Age of the Cyborg Advisor

My Friend "Robbie the Robot"

In the July 31, 2017, issue of *Barron's* magazine, Alex Eule reports on robo-advisors' competitive landscape from the pioneer Betterment, the self-described IA replacement automation, to the large and late entrants (Eule, 2017). Betterment, founded in 2010, is the largest independent RIA, with $9 billion AUM, but that's small compared to 2015 entrants Schwab ($19 billion) and Vanguard ($83 billion). Clearly, the large firms are where the AUM is housed already, so it has been easier to convince clients to switch to lower-cost digital portfolios. Confirming what we present, and despite the media conviction that Millennials are perfect for digital advice, the robo harvested assets are coming from older and wealthier age groups. As Alex Eule of Barron's points out, "Eighty percent of Vanguard Group's digital clients, for example, are above the age of 50. Betterment skews younger, but the bulk of assets and accounts still comes from the 35-to-55 age range." Why? Because that is where the money is. And as a sign of industry direction, Betterment is adding human advisors for higher tier AUM accounts (Fig. 3.6).

Much of the research for the *Barron's* article comes from BackendBenchmarking.com, itself founded by an independent wealth management firm, Condor Capital ($1 billion AUM). Backend opened and funded 16 different robo-advisor accounts to see how they performed. Schwab led the pack with 11.94% returns over the year. More important than performance, the final comments from Vanguard and Betterment indicate that Backend does not "… pick up the nuances of an advisory relationship…The nuances are what Vanguard calls advisor alpha – which it calculates can add about 3% to the returns of advisor-managed portfolios" (Winokur Munk, 2017).

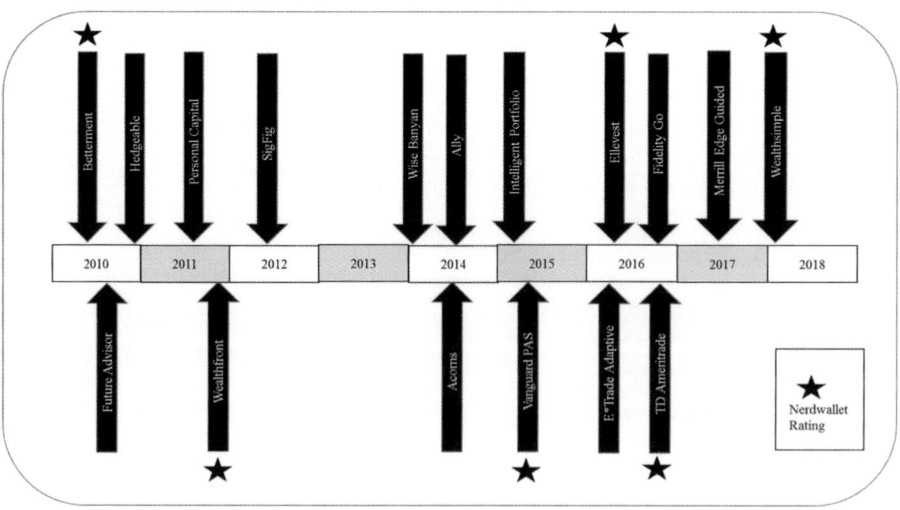

Fig. 3.6 Over 40 wealth management apps

Tools of the Advising Trade

Envestnet is the largest provider of wealth management tools to over 50,000 advisors with excess of $1 trillion in assets on their platform. Envestnet did not build, but instead bought Yodlee for $590 million to gain access to technology for advisors, most of whom are at independent firms (Malito, 2015). As Bill Crager, president of Envestnet, has said, "It's what has been missing." This acquisition allowed advisors' clients to link their bank and credit card information to their advisors' platform, which is essential for the holistic type of advice we are discussing. Based on the data aggregation and analytics it acquired in 2015, Envestnet is pulling data from hundreds of thousands of accounts at more than 15,000 financial institutions and sifting through billions of transactions a day (Dodds, 2017). Advisors can see assets held away and can also review client's spending habits, credit card, or mortgage debt, which overall helps advisors provide better advice. Over the next few years, the services will include artificial intelligence tools, voice functionality (think Alexa or Siri), and other apps that will make it easier for clients and advisors to communicate. Envestnet is not attempting to replace the advisor, but simply continuing to do what it has always done, help the advisor (Arora, 2015).

Envestnet has continued its acquisitions with the announcement in September 2017 of the proposed purchase of FolioDynamix, a competitor with more than 3.2 million client accounts and over $800 billion in assets. When complete, Envestnet will be working with almost $2 trillion in assets and approximately 10 million investor accounts. Blackrock, in partnership with InvestmentNews Research, studies the "Elite RIA," which they define as the most successful firms in the financial advisory industry. The 2017 study shows, as in past years, the Elite RIAs getting larger, gathering more assets, and becoming an even stronger force, not only in the independent space but in the entire financial services industry (InvestmentNews Research, 2015). They are "implementing broader fee offerings with more service offerings, relying on technology to deliver those services and smooth the back-office processes." Morgan Stanley, as stated previously, has begun employing machine learning to match investment possibilities to client preferences. Envestnet, Blackrock, and Morgan Stanley, as examples, all help confirm our hypothesis. Clients and advisors want technology to assist (not replace) the human relationship (Davenport and Bean, 2017).

Embracing Change: Advisor and Client Need Alignment

The human touch is where the relationship is formed that feeds client literacy and advisor insights. The human touch is where the new alpha can be found, not by myopic portfolio stock picking but through an integrated view of the client's goals and delivering on those goals. A tall order for a profession challenged to maintain professional literacy while advising increasingly illiterate clients in a rapidly changing industry.

One way to create literacy and improve competency is by creating a team of experts, each knowledgeable in specific areas, since the data and complexity are too large for optimal results to be expected from a single individual advisor armed with a Bloomberg terminal and an Excel spreadsheet. Plus, the regulators strongly frown on advice being offered from those without the formal training and designations. We believe a CFP can be the quarterback, with a general knowledge of all areas, supported by a team comprised of a CPA (tax), lawyer (wills, estates), insurance agent (risk management), financial advisor (investments), and behavioral economist (goal transitions). There is overlap within the team, any of the above could be a CFP and/or a CFA. Other significant designations only increase both the literacy and competence of the team and raise the quality of the service. Over this decade of industry team formation, we expect new blended positions to be invented as standard team positions merge and expand anew.

The "tool kit" is a good image to reference with a balance between investments and the specific ever-changing goals of each client. Half of the value proposition is addressed using a tool kit and a team of literate experts managing the portfolio through market changes. The second half of offering life-cycle goals advice, in addition to portfolio composition, means reviewing all planning needs: insurance, estate, retirement, and education planning. The first challenge is truly analyzing the advising/client process and identifying where tools can help connect the team. The second will be integrating behavioral economics into services during client goal revisions and decision-making.

A good starting point and a great tool kit for "Knowing Your Customer" was created by and for the Certified Financial Planning (CFP) industry, and it is known as the EGADIM model (Fig. 3.7). EGADIM stands for Establish relationship, Gather data, Analyze data, Develop (and communicate) the plan, Implement the plan, and Monitor the plan. The major regulatory agencies' assessment models follow a similar approach for the construction and maintenance of a compliant business. Financial technology (FinTech) startups are now targeting every step of EGADIM. Today's computer automation solutions target client information enrichment by mining Twitter, Facebook, and LinkedIn, to deliver a more personalized profile of the client. The assembled profiles are pushed to customer relationship management (CRM) software, such as Salesforce and Redtail Technology, to drive credit rating bots, as well as the human advisor's cold call. But today's analytics require specific client data or the applied analytic produces solutions that are simply noise. Human clients are not going to sit at a screen or speak to a bot and answer a mind-numbing number of profile questions, estimated today to be over 400 for initial account opening. A human advisor is needed to shape the profile data with emotional intelligence about the relationship. Automated data gathering via a "bot" is impressive in the details that can be found and equally concerning in the facts that can be missed including Know Your Customer (KYC) and anti-money laundering (AML) regulatory reporting. Once qualitatively valued and curated for applicability, both the "bots" and the humans use the same analytic tools to analyze, develop, and implement product and scenario plans for the assets. The "bots" are faster and produce fewer errors in creat-

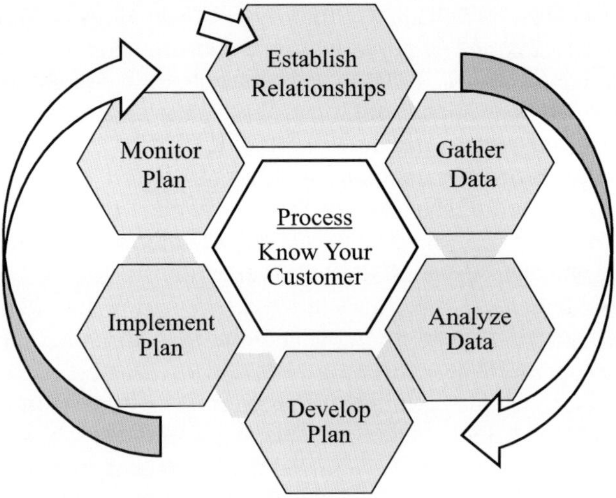

Fig. 3.7 Know your customer process

ing multiple strategies. However, it is the humans that meet with clients, communicate the choices, and verify understanding and agreement.

Change is about abandoning old methods and models and adopting new ones. The existing vendor solutions will be forced to change or lose markets to new FinTechs focused on disaggregated processing steps. As more client-centric services are provided, just adding functionality to existing heirloom systems will only increase the cost of ownership, without the added benefits of an enriched client profile linked to specific client life goals and priorities.

Customer relationship management software, such as Salesforce.com, can cost over \$10,000 for initial setup and \$15,000 a year for a firm with just six users. Custodians, who house client assets, send out statements, and in most cases execute and process trades, rather than charge flat fees, usually have trading minimums that could be \$100,000 a month, plus extra costs for additional services and reports. Costs vary, and custodians focused on services for independent advisors, such as Schwab, Fidelity, and TD Ameritrade, charge less but offer less sophistication. Companies such as Envestnet also have some initial setup fees, as well as ongoing fees ranging from 8 basis points (0.08%) up to 50 basis points, plus they also have minimums, upward of \$25,000 a quarter. Lastly, on the regulatory side, Erado for email monitoring and archiving, NCS for third-party compliance, Quest CE for mandatory continuing education, and RIA in a Box for compliance and regulation could total tens of thousands of added expenses. And you still need a team that can use the tools and can convert client goals into plays, putting points on your balance sheet.

The regulatory rules are standardizing the operational process and the tool box also has to comply with the rules of play. The simplest regulatory rule to focus on is KYC (FINRA rule 2111). The regulators will continue to drive compliance officers to know as much as possible before authorizing advisors to offer services, but how does a compliance officer evaluate a robo-advisor? Technology changes the historical governance process and requires advisor licensing and precedence case law review. That is change with a capital C, all the way to the C-suite of the large firms with the unanswered question of liability weakly defined (Sheldon v. SEC, 1995).

Regulators still write the rules, and the SEC has made it very clear that "ultimately the responsibility for a broker-dealer's compliance resides with its chief executive officer and senior management." On the registered RIA side, the SEC puts the ultimate responsibility on "senior management," without clearly indicating what that means. The SEC has also expressed concern with outsourcing the role of chief compliance officer and has stated in Rule 206(4)-7 that "each advisor registered with the (SEC) designate a chief compliance officer to administer its compliance policies and procedures." The bottom line is that regulators are fine with firms using technology to supervise, but the ultimate responsibility still lies with human owners of the business (SEC, 2015).

The courts, not the regulators, have forced the last area of change: the protection of client information in addition to assets. This area has almost always required advisors and brokers to custody client assets at a third-party firm, such as JP Morgan, Pershing (Bank of NY), Fidelity, Schwab, and TD Ameritrade, to name the biggest. The cost of their custody and clearing services range from 10% to 20% a year, depending on the types of services needed. These fees have stayed stable as the financial service industry continues to recover from the 2008 financial recession, but the lack of real competition and the large barrier to entry leads us to believe that custody of client assets and information will become an even larger expense in the future, especially for the smaller firms. CEOs and CCOs within the industry identify compliance is their largest expense without a doubt. In a study done by IAA and ACA Compliance, 48% of IA firms spend between $100,000 and $500,000 annually, while 14% spend between $1 million and $5 million. The access to historical data is required to train the robots using machine learning, yet most custody firms are just beginning to offer advisors any ability to export client history. The use of cloud-based solutions will help with standardized application program interfaces (API) to move the data, but privacy or security is weakened (Careiro & Lamba, 2016).

One thing is certain: technology is here to stay. Services, such as those mentioned above, are required by regulators, expected by clients, and desired by advisors. But they cost a lot of money, and, when combined with fee compression, make it that much harder for small, independent advisory firms to survive.

The Nimble Survive

Fiduciary Role

The smaller independent advisory firms are facing many issues, which can be turned into opportunities. Fee compression due to passive management and robo advising is forcing a team-based approach. But changing demographics was forcing advisors to create teams to advise clients on all of life's financial decisions anyway. Even as big firms, like Morgan Stanley, augment the advisor with technology, it is the human advisor (not the robot) calling to wish the client happy birthday and building a trust relationship. Informed personal knowledge of the client and a suite of analytic functionality is exactly what both the Boomer and the Millennial want in a human IA the opposite of a cloud delivered AI (artificial intelligence). The Boomers, as they are living longer with more complicated needs, demand the team has a CPA, estate lawyer, and investment professional all on the same page. The Millennials may not know exactly what they want, but as time goes on, we believe personal, unbiased, customized advice will be in demand for Generation X, Generation Y, Generation Z, and beyond. Regardless of generation, goal-based financial decision-making is the answer. These very different generations are moving away from "following the market" and focusing on an ever-changing financial plan that meets each client's unique needs. Technology can augment both sides of the relationship increasing literacy in the client as well as increasing advisor's scope and AUM.

Regulators are also looking at the process and requiring, in many aspects, a fiduciary relationship. Simply put, that is putting the client's needs ahead of the advisor and firm. It is difficult for robo advising to act as a fiduciary. Confirming this, Betterment announced on July 26, 2017, that they were rolling out fiduciary financial advice to most of its 280,000 clients without raising its price. To act as fiduciary is the highest standard of financial advice, which robo-advisors are providing at fees ranging from 25 to 40 BPs today (going to 3 BPs tomorrow). Betterment's announcement is confirmation that human advice is alive and well and further indication that robo advising functionality and focus continue to mature.

Trust relationships are the natural advantage of the human advisor team. Clients have witnessed Madoff, Wells Fargo, and the Great Recession, and as such, the advisory industry needs to work hard to regain trust. There is a tremendous opportunity to use technology for cost savings and greater operational efficiencies to support a team. In a recent report from Financial Technologies Forum, Optimas forecasts a roughly 75% increase in spending by 2021 to invest in IA augmentation. Economies of scale are not essential for firms with less than $1 billion in assets under management to survive. But these firms must build on the strengths that have already made them successful, while responding to increased functional requirements. Asset gatherer tools are readily available, as well as tools to improve asset management. Becoming a better asset manager today requires using computer-automated investing tools or wealth management platforms like Envestnet to increase productivity. The breakeven for these services is around $150 million in assets under management (AUM), and the technology will get less expensive as companies like Betterment, Blackrock, and possibly Google and Amazon compete for advisory business.

Attracting New Entry-Level Advisors

For literacy and competence, as well as succession planning, a Certified Financial Planning holistic offering will attract more clients (young and old) and make the industry more attractive to advisors (young and old). By joining a team, the entry-level advisor does not have the pressure of seeking out assets but can learn and be mentored by the experienced professionals, which eventually solves the succession planning dilemma. Fortunately, the CFP roadmap is being taught nationwide in almost 400 registered programs. These CFP programs and licensing, including continuing education, help build a base of financial professionals delivering on their fiduciary duty. Additionally, there are jobs available right now for those with the CFP designation (Malito, 2016).

We think the key is for the older, better established advisor to create apprenticeships and internships with universities in their community. University programs, such as student-managed investment funds, can and should be developed with an objective beyond simply managing money. The universities need to foster these programs to get students practical experience to complement their theoretical knowledge. Upon graduation, these students can become full-time employees and need to be brought into client meetings. On the client side, parents need to make sure their children know what decisions are being made. The next generation of client and advisor need to meet and become a part of the conversation. Even if Millennials look at financial decision-making differently than Generation X or Boomers, they have goals, and it is the advisor's job to help them meet those goals. As the data shows, the money is in motion, and eventually the next generation will be responsible for it.

The trends continue to let the assets gather at large financial firms (JP Morgan, Morgan Stanley, etc.), large asset managers (Blackrock, Vanguard, etc.), and large custodians (Pershing, Schwab, etc.). While this is fine for now, the advisors need to be careful because it is difficult to move money and data from those places. The independent advisor needs to maintain their independence. It is essential if they are going to show client's strategies and products that best meet their goals, rather than what the wirehouse requires them to sell. The ability to offer multiple types of products, both traditional and alternative, is the best differentiator between large and small firms. The best protection the advisor has is to maintain strong client relationships, while beginning to build those same ties with the next generation.

Closing Comments

As digital transformation overlaps with generational wealth transfer, the way forward is easier said than done, but there is time, the tools are available with more on the way, the money is in motion, and the clients need the help. Only with technology augmentation of a multidimensional service team will the client find an increasing value proposition. Simply put, revenue will come from the human touch, not the machine screen, found in technology augmented team. A team with technology will always beat a team without technology, and a team with technology will always beat a technology with no team.

We conclude and summarize with the following three assessments of the RIA industry today:

1. The information age is on the ascent, and just as the industrial age showed no mercy to the Luddites, so too must IAs adapt, adopt, and learn.
2. The information age can improve client literacy if the IA shapes the digital conversation and empowers the client as wealth is inherited or accumulated.
3. More client-specific custom service requires an enriched multi-disciplined team with the technology augmentation to service the client's life-cycle timing while matching near-term goals.

References

Accenture. (2015). The greater wealth transfer: Capitalizing on the intergenerational shift in wealth. *Accenture*. Retrieved from https://www.accenture.com/t20160505T020205Z__w__/us-en/_acnmedia/PDF-16/Accenture-CM-AWAMS-Wealth-Transfer-Final-June2012-Web-Version.pdf#zoom=50

Armstrong, D. (2017, June 29). *Advice industry shrinking, even as fees drop*. Retrieved from http://www.wealthmanagement.com/industry/advice-industry-shrinking-even-fees-drop

Arora, A. (2015, August). *Yodlee Financial Cloud Company overview*. Retrieved from https://www.sec.gov/Archives/edgar/data/1161315/000119312515285657/d30062d425.htm

Baby Boomers will control 70% of disposable income. (2016, February 22). Retrieved from https://impactbp.com/baby-boomers

Barr, K., & Gebauer, J. (2016). *A profile of the investment advisor profession* [2016 Evolution Revolution].

Benjamin, J. (2017, February 22). *Beware the gap coming in the number of financial advice professionals*. Retrieved from http://www.investmentnews.com/article/20170222/FREE/170229971/beware-the-gap-coming-in-the-number-of-financial-advice-professionals

Berridge, S. (2014, September 9). *Millennials after the Great Recession*. Retrieved from https://www.bing.com/cr?IG=90ADE2A85F3040E58582CB1B20F0B4E0&CID=089528814C376278278423154D98634B&rd=1&h=ShMEpn60JDPUOTXSFAwjWIsXNX7561OWR3NvLiI8K5E&v=1&r=https%3a%2f%2fwww.bls.gov%2fopub%2fmlr%2f2014%2fbeyond-bls%2fpdf%2fmillennials-after-the-great-recession.pdf&p=DevEx,5067.1

Bertolucci, J. (2013, December 31). *Big data analytics: descriptive vs. predictive vs prescriptive*. Retrieved from https://www.bing.com/cr?IG=C30C00C0E35941A19BF1B52246169D83&CID=2B0AF7F2253368D3184DFC66249C69A8&rd=1&h=p6hR6L9_n893dK7cNO4pRdu5zxGNhxIEeCnSIpS5n9k&v=1&r=https%3a%2f%2fwww.informationweek.com%2fbig-data%2fbig-data-analytics%2fbig-data-analytics-descriptive-vs-predictive-vs-prescriptive%2fd%2fd-id%2f1113279&p=DevEx,5069.1

Britton, B. L., & Atkinson, D. G. (2017, March 01). *An investigation into the significant impacts of automation in asset management*. Retrieved from https://www.questia.com/library/journal/1P3-4321313575/an-investigation-into-the-significant-impacts-of-automation

Brudner, E. (2016, February 23). *6 Data-driven reasons why cold calling flat out sucks*. Retrieved from https://blog.hubspot.com/sales/cold-calling-flat-out-sucks

Bureau of Labor Statistics - Occupational Outlook Handbook – What Personal Financial Advisors Do. (2018, April 13). Retrieved from http://www.bls.gov/ooh/business-and-financial/personal-financial-advisors.htm#tab-2

Carreiro, L., & Lamba, S. (2016, April 27). *2016 Investment management compliance testing survey*. Retrieved from https://www.acacompliancegroup.com/news/compliance-alert/2016-investment-management-compliance-testing-survey

Collins, J. M. (2012). Financial advice: A substitute for financial literacy? *SSRN Electronic Journal*, 307–322. https://doi.org/10.2139/ssrn.2046227

CFP Board. (2018). *CFP certification: The standard of excellence*. Retrieved from https://www.cfp.net/about-cfp-board/cfp-certification-the-standard-of-excellence

Cornfield, J. (2016, July 06). *Only 1 in 3 millennials are investing in the stock market*. Retrieved from https://www.prnewswire.com/news-releases/only-1-in-3-millennials-are-investing-in-the-stock-market-300293359.html

Davenport, T. H., & Bean, R. (2017, August 03). *How machine learning is helping Morgan Stanley better understand client needs*. Retrieved from https://hbr.org/2017/08/how-machine-learning-is-helping-morgan-stanley-better-understand-client-needs

Dodds, L. (2017, March 24). *Financial services to lose 230,000 jobs to AI systems*. Retrieved from https://www.ftfnews.com/financial-services-to-lose-230000-jobs-to-ai-systems/16683

Eule, A. (2017, July 29). *Rating the robo-advisors*. Retrieved from https://www.barrons.com/articles/rating-the-robo-advisors-1501303316

Fava, M., Boersema, J. & Hamalolu, U. (2016). EY – The next generation of financial advisors. Ernst & Young LLP Wealth Management

Fiegerman, S. (2017, September 8). *The biggest data breaches ever.* Retrieved from http://money.cnn.com/2017/09/07/technology/business/biggest-breaches-ever/index.html

Fry, R. (2016, April 25). *Millennials overtake Baby Boomers as America's largest generation.* Retrieved from http://www.pewresearch.org/fact-tank/2016/04/25/millennials-overtake-baby-boomers/

Harding, A., & Allison, K. (2016, April). *Employee financial wellness survey* [Survey]. PwC, UK.

Haslem, J. A. (2008). Why do mutual fund investors employ financial advisors? *SSRN Electronic Journal*. https://doi.org/10.2139/ssrn.1115886

InvestmentNews Research, BlackRock launch new study of … (2015, May 28). *InvestmentNews*. Retrieved from http://www.bing.com/cr?IG=ADA75485406446C28C480447C53123A0&CID=2B43F1CA1E446A352537FA5E1FEB6BC9&rd=1&h=ck1mc_uYJPHCWnrdUGm-CL8vnCXoFQM-bGTLb45_Jpxw&v=1&r=http%3a%2f%2fwww.investmentnews.com-%2farticle%2f20150528%2fFREE%2f150529930%2finvestmentnews-research-blackrock-launch-new-study-of-elite-rias&p=DevEx,5069.1

Jain, A., Hyde, P., & Lyman, S. (2013, October 04). *Wealthy, young, and ambitious: How banks can profitably serve the rising mass affluent*. Retrieved from https://www.strategyand.pwc.com/reports/wealthy-young-and-ambitious

Kitces, M. E. (2016). The future of financial planning in the digital age. *CFA Institute Conference Proceedings Quarterly, 33*(2), 17–22. https://doi.org/10.2469/cp.v33.n2.1

Levine, A. (2017, Spring). Merrill Edge report. Retrieved from https://www.merrilledge.com/

Lustig, R. H. (2017). *The hacking of the American mind: The science behind the corporate takeover of our bodies and brains*. Penguin Random House. ISBN 9781101982587

Malito, A. (2015, August 23). Envestnet is the industry's 800-pound tech gorilla. *Investment News*. Retrieved from http://www.investmentnews.com/article/20150823/REG/308239997/envestnet-is-the-industrys-800-pound-tech-gorilla

Malito, A. (2016, May 29). The time is now for young financial advisers. *Investment News*. Retrieved from http://www.investmentnews.com/article/20160529/FREE/160529937/the-time-is-now-for-young-financial-advisers

Mellan, O., & Christie, S. (2017, February 27). Thinking millennial: How to woo the largest generation. Retrieved from http://www.bing.com/cr?IG=31E9DDAE1E884B068731B9BA3F111A87&CID=21B93B4526CC61B33DE130D127636097&rd=1&h=lObjrl6OcoNB6ZwineQQp-3IIw9yVOdY2v3NXJIdj_c&v=1&r=http%3a%2f%2fwww.thinkadvisor.

com%2f2017%2f02%2f27%2fthinking-millennial-how-to-woo-the-largest-generat&p=DevEx,5054.1

Pirker, A., & Schmitt, S. (n.d.). Building a wealth management practice: Measuring CFP professionals' contribution. Retrieved February 12, 2018. Aite Group, LLC.

SEC Issues Risk Alert on Outsourced CCO Model. (2015, November 16). Retrieved from https://www.acacompliancegroup.com/news/compliance-alert/sec-issues-risk-alert-outsourced-cco-model

Shekhtman, L. (2016, July 06). *Why so few millennials invest in the stock market*. Retrieved February 12, 2018, from http://www.businessinsider.com/why-so-few-millennials-invest-in-the-stock-market-2016-7

Sheldon v. SEC, 45 F.3d 1515, 1517 (11th Cir. 1995). The president of a corporate broker-dealer is responsible for compliance with all of the requirements imposed on his firm unless and until he reasonably delegates particular functions to another person in that firm, and neither knows nor has reason to know that such person's performance is deficient.

Son, H. (2017, May 31). *Morgan Stanley's 16,000 human brokers get algorithmic makeover*. Retrieved from https://www.bloomberg.com/news/articles/2017-05-31/morgan-stanley-s-16-000-human-brokers-get-algorithmic-makeover

Southall, B. (2017, July 28). *Scott Hanson sells RIA to Parthenon after 'semi-retirement' drove him half-crazy – so he could work full-time to get to $10 billion*. Retrieved from https://riabiz.com/a/2017/7/28/scott-hanson-sells-ria-to-parthenon-after-semi-retirement-drove-him-half-crazy-so-he-could-work-full-time-to-get-to-10-billion

Stich, A. (2017, September 07). *Winning the loyal business of Gen X with technology*. Retrieved from https://www.linkedin.com/pulse/winning-loyal-business-gen-x-technology-anthony-m-stich

Tedesco, D. (2015). I, robo-adviser? Creating the blended adviser experience. *Journal of Financial Planning, 28*(1), 17.

Tibergien, M., & Dellarocca, K. (2017). *The enduring advisory firm: How to serve your clients more effectively and operate more efficiently* (1st ed.). New York: Wiley/Bloomberg Financial.

Touryalai, H. (2012, August 08). One of the fastest-growing careers is in desperate need of young talent. *Forbes/Investing*. Retrieved from https://www.forbes.com/sites/halahtouryalai/2012/08/08/one-of-the-fastest-growing-careers-is-in-desperate-need-of-young-talent/#6d1449ee53c6

Wharton Executive Education. (2018). *Active vs. passive investing: Which approach offers better returns?* Retrieved from http://executiveeducation.wharton.upenn.edu/thought-leadership/wharton-wealth-management-initiative/wmi-thought-leadership/active-vs-passive-investing-which-approach-offers-better-returns

Winokur Munk, C. (2017, September 15). The changing landscape for independent advisors. *Barrons*. Retrieved from https://www.barrons.com/articles/the-changing-landscape-for-independent-advisors-1505533992

Zulz, E. (2017, May 4). *BlackRock's Fink: Tech to overtake its money managers soon*. Retrieved from http://www.thinkadvisor.com/2017/05/04/blackrocks-fink-tech-to-overtake-its-money-manager

Chapter 4
Improving Fleet Management Strategy and Operational Intelligence with Predictive Analytics

Bill Powell and Suresh Chandran

Abstract The fleet management industry is comprised of fleet management providers that help ensure an organization's vehicles remain on the road supporting core business functions efficiently, safely, and at the lowest total operating cost. Recent technology advancements in the data analysis space coupled with enriched data domain have made it possible for analytics to be applied strategically for fleet management solutions. One of the latest and game changing services to enter the fleet management market in recent years is in the IoT (Internet of Things) space, specifically, vehicle telematics services. Collating pure telematics information with other information from other areas such as maintenance, fuel, and driver performance can improve fleet management. Using the case study of ARI, this paper explores the application of advanced analytics to various facets of fleet management and ARI's experience in aligning analytics with its business strategy. The paper also outlines the steps needed to implement a telematics and analytics strategy in organizations and the importance of bridging the gap between theory and practice.

Keywords Fleet management industry · Data · Analytics · Telematics · Internet of Things · Operational intelligence · Business strategy

Introduction

Fleet managers face increasing economic challenges in the acquisition, operation, and disposition of their company cars and trucks (PWC, 2015), especially in vocational fleets where complex work trucks and equipment are vital to their organization. The acquisition and configuration of company assets (cars and trucks) have

B. Powell
ARI, Mount Laurel, NJ, USA
e-mail: bpowell@arifleet.com

S. Chandran (✉)
Drexel University, Philadelphia, PA, USA
e-mail: sc44@drexel.edu

M. Anandarajan, T. D. Harrison (eds.), *Aligning Business Strategies and Analytics*, Advances in Analytics and Data Science 1,
https://doi.org/10.1007/978-3-319-93299-6_4

become ever more intricate due to evolving options and expanded supply chain processes. Once the asset is built, delivered, and put into service, the fleet managers' attention turns toward the effective operation of the fleet. Decisions regarding fuel management, vehicle maintenance, and asset uptime can have a profound effect on the fleet's total cost of ownership. Finally, once the asset has reached its economic breakeven point, multiple options need to be considered to maximize its residual value. Thus, smart and strategic business decisions must be made throughout the entire life cycle of the fleet from acquisition to operation to disposal. Furthermore, it is essential that these choices are balanced with the safe and effective use of the asset.

Recent technology advancements in the data analysis space coupled with enriched data domains have made it possible for analytics to be applied strategically for fleet management solutions. One of the latest and game changing services to enter the fleet management market in recent years is in the IoT (Internet of Things) space, specifically, vehicle telematics services. Telematics is a technology used to interface with, obtain, and measure diagnostic and positional data from moving assets (cars, trucks, trains, etc.) that can be analyzed and leveraged to make better decisions about their location, type of use, status, and behavior. GPS data provide the exact location and speed of the asset, while numerous vehicle sensors capture its idle time, battery voltage, fuel/air ratio, carbon output, and other performance metrics including any diagnostic issues should they arise. This information is streamed at high velocity over a cellular network to a centralized facility where it is stored and combined with other empirical data (fuel, maintenance, accident information, etc.) to form a more complete picture of how the asset is operating. Telematics devices are either plugged into the asset's OBD-II port (onboard diagnostic port) or directly wired into the asset depending on the vendor, vehicle manufacturer, and required capabilities.

The use cases for pure telematics information are vast. However, the real potential for this information is when it is collated with other data domains such as maintenance, fuel, and driver performance and aligned to the fleet management strategy. Our paper explores applications of advanced analytics in various facets of fleet management and demonstrates these applications through ARI's experience in aligning analytics with its business strategy.

Background

The Fleet Management Industry: An Overview

The fleet management industry is comprised of fleet management companies that are partnered with organizations to help manage the vehicles they need to conduct day-to-day operations as a strategic business asset that contributes to the bottom line. Fleet management companies help clients by optimizing the entire lifecycle of the vehicle beginning with the acquisition of the car or truck, continuing to programs and services such as maintenance management and licensing compliance and through to the eventual sale of the vehicle. Fleet management providers help ensure that an organization's vehicles remain on the road to support core business functions efficiently, safely, and at the lowest total operating cost.

The fleet management market is constantly evolving, and the richness of telematics data has provided insights that previously were difficult to obtain or riddled with inaccuracies. Telematics enables more accurate odometer readings, very precise vehicle positional information, and the ability to determine how the vehicle is operated. Drivers who have a propensity to speed are now easily identified in addition to those who tend to start and idle their vehicles for an excessive amount of time (each hour of idling consumes about one gallon of fuel). This information can be used to help customers slow down their fleet, which improves fuel economy. In other cases, customers have provided their insurance companies with details about their fleet's performance to negotiate lower rates. Another area where geospatial information has helped reduce costs is for customers who operate Department of Transportation (DOT)- compliant vehicles. Customers who operate a DOT vehicle on a federal highway are required to pay a tax for the distance traveled. Typically, customers take a conservative approach and overpay. However, if the asset has telematics capabilities, customers can be assured they pay only for the exact distance traveled.

Similar to the DOT use case, drivers who operate a company vehicle and are allowed to use it for personal use must declare their personal mileage as a fringe benefit to the IRS. The process of recording business and personal mileage is tedious and error prone. To streamline the process, automatic recording for vehicles equipped with telematics was recently introduced. By geocoding the driver's home and office, the system can distinguish business trips from personal use. Drivers can now simply skim and verify the information before it is posted to their tax department instead of manually keeping detailed driving logs.

Aligning Fleet Management Strategy with Analytics: The ARI Case Study

ARI is the largest privately held family-owned fleet management company in the world. It provides an array of products and services specifically designed to control and reduce its customers' fleet expense, increase the fleet's availability, and allow the fleet to be operated in a safe and effective manner. Over 1.4 million vehicles are managed by ARI across the globe with large concentrations in North America and Europe. Focusing on complex vocational fleets is a key component of ARI's strategy. ARI provides a full range of services including acquisition, supply chain management, maintenance, telematics, licensing and compliance, fuel management, accident claims and subrogation, driver safety, and vehicle remarketing. Each of these services generates volumes of information; this is collected, collated, and analyzed to understand their underlying factors and provide recommendations for continual cost reductions and improvements in efficiency.

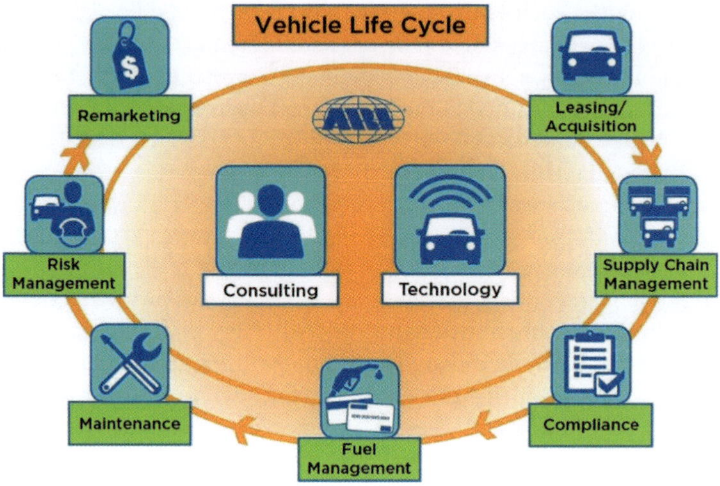

Fig. 4.1 Vehicle life cycle management

The fleet management lifecycle illustrated in Fig. 4.1 consists of numerous processes and decisions to control the asset's cost and ensure its reliability categorized into several major subject areas: acquisition, supply chain management, compliance, fuel management, maintenance management, risk management, and remarketing. The acquisition and supply chain phases entail aligning the customer's current and future needs with the best options based on the manufacturer's availability. ARI uses status feeds from its vendor network to continuously monitor the assets as they are built to ensure timely delivery and to take proactive measures if disruptions in the supply chain are detected.

Once the asset is built, delivered, and put into service, ARI provides various operational services including fuel, maintenance, and risk management. ARI receives electronic feeds from its fuel vendors for each transaction that occurs. The data are very granular and include the type of fuel purchased (diesel, premium unleaded, regular unleaded, etc.), the gallons/liters purchased, the cost per unit, and the total cost for the transaction. Analytics are used for both audits and cost control to identify anomalies (pilferage) and areas of opportunities (the best station to use based on proximity and price). Maintenance management is aimed at maximizing the asset's uptime/availability through preventive maintenance and negotiations about repair costs. By using historical repair information combined with vehicle sensor data (IoT) to predict future failures, the appropriate proactive correction can be determined to minimize major component failures.

Risk management consists of an array of services aimed at improving driver safety. Data collected from motor vehicle reports and IoT sensor information are combined with empirical data captured from fuel and maintenance transactions to form a comprehensive picture of how the driver operates the asset (speeding, aggressive behavior, consistently late for preventative maintenance service, etc.). Analytics promote this service through personalized recommendations based on enriched data and enhanced statistical analyses. Fleet management companies can intercede and assign specific driver safety modules and courses that are relevant to the driver's propensities (to speed, drive aggressively, etc.) with the objective of changing driver behavior before an issue or accident occurs. Such proactive actions lead to safer driver behavior and longer-term cost savings. The Red Hawk Fire & Security LLC use case described later in this article is an example of this aspect of fleet management.

Once the asset has reached its economic breakeven point, the inflection point where it costs more to maintain the vehicle than it is worth, ARI's remarketing service begins the process of selling the unit via upstream and downstream options in the timeliest manner for the best margin. Upstream opportunities include selling the asset to the driver or making it available for purchase by other employees of the customer. This method typically yields the quickest results and is seen as a perk for the customer's employees. Downstream methods include consignment sales, BuyDirect – a program where ARI purchases the vehicles and assumes the resale risk – and selling the asset through physical and virtual auctions. Information from multiple auction sources is analyzed in real time to determine the optimal time, price, and location to sell each asset. ARI sells over 80,000 assets annually. It can quickly provide its subject matter experts with insights into market demand in an easily consumable manner to maximize value for its customers.

ARI's telematics strategy depicted in Fig. 4.3 is squarely focused on obtaining and leveraging information from devices. As such, its approach is to remain device and vendor agnostic. Data integration is a critical component of this strategy. Over the last few years, ARI has preintegrated itself with over 20 telematics vendors and has made significant capital investments in its analytic capabilities. Understanding that many of ARI's customers operate in diverse geographical areas, customers may need more than one telematics vendor to cover their territory. ARI's vendor agnostic approach has allowed the company to integrate, collate, and analyze data from disparate telematics vendors. Previously, customers using multiple providers would log in to each vendor's site, download the applicable data, and stitch the analyses together themselves. Certain fleet customers may use multiple telematics vendors due to the composition of their fleet (cars, light duty trucks, heavy duty trucks, equipment, etc.), geographical challenges, or simply because they are required to use multiple vendors instead of single sourcing. The agnostic strategy removes the customers' multi-vendor integration concerns and processes, provides a single system for all of their telematics data, and emphasizes deriving value from information instead of the complexities of obtaining and managing it.

Recent estimates project that the global telematics market will grow exponentially and anticipate that approximately 105 million new cars will have some form of connectivity by 2025. Companies looking to improve efficiencies and introduce a comprehensive driver safety program are investing in telematics. The granular and near real-time data can create opportunities for pay-as-you-play types of insurance programs, rewarding safe drivers and companies with lower premiums. Positional information garnered from the vehicle's telematics device can help the logistics process tremendously by providing live updates of the asset's location with

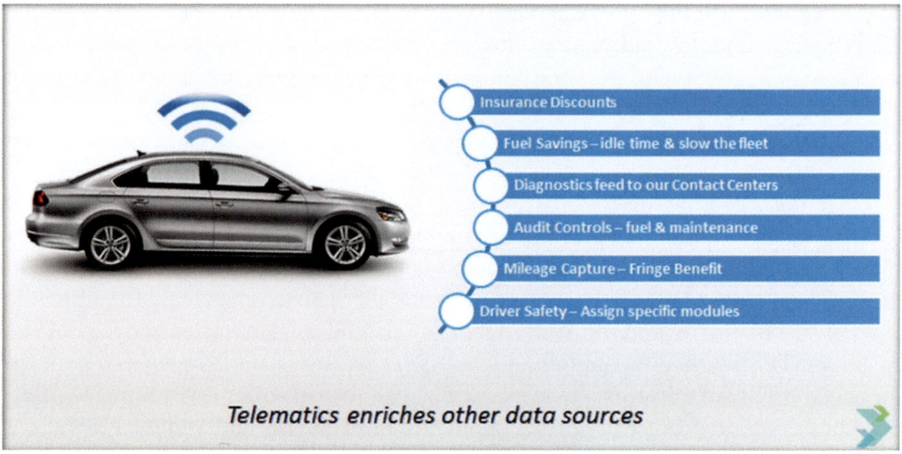

Fig. 4.2 Telematics framework: IoT in fleet management at ARI

intelligent routing to avoid congestion and areas prone to accidents. As the technology continues to mature and become further integrated, two-way capabilities will begin to emerge. However, this level of sophistication will likely be reserved for the manufacturer. A recent example is Tesla's over-the-air reprogramming of select vehicles in the path of Hurricane Irma. Tesla's software update extended the battery's capacity, allowing its customers greater driving range to avoid and escape Irma's path, all without the need to visit a service facility (Fig. 4.2).

As mentioned previously, the potential for telematics is better unleashed when combined with key performance indicators related to maintenance, fuel, and driver performance. The appropriate key performance indicators are determined during a discovery phase with customers to understand their business needs and the items most relevant to them. Delivering key performance indicators in the proper context is an important part of the process if the information is to be actionable. As an example, if the customer is a CFO, his/her needs will focus on fiscal analysis and actual-to-budget items as opposed to a fleet manager whose key performance indicators will relate to maximizing operational performance. ARI provides recommended pre-built dashboards based on customer preferences, which can be further tailored to match their specific requirements. ARI's experience aligning its services with analytics is documented below.

Aligning Maintenance and Vendor Management Strategy with Analytics

ARI's maintenance management program consists of multiple call centers with automotive service excellence (ASE)-certified technicians who interact with vendors to manage asset repairs on behalf of its customers. Knowing which components need replacing and obtaining the best price for the part and labor has traditionally been

Fig. 4.3 Key performance indicators in maintenance analytics

learned through experience. Decisions relating to repair vs. replace, which vendor the driver should be referred to, and if the issue is under the manufacturer's warranty all need to be made accurately and efficiently in an environment where time is of the essence. A new system called VMS (vendor management system) has recently been created to reduce the number of inbound calls into the call center and increase the overall customer and vendor service experience. Vendors can request the recommended repairs via a new website that is based on an advanced rule-based engine. Under the cover are advanced analytics that use an array of information to determine if the repair is needed, if the cost is within acceptable tolerances based on previous repairs and location, and if the repair needs to be handled by an ARI technician.

Maximizing asset uptime and ensuring that the fleet is available are critical for ARI's customers. By leveraging predictive algorithms to determine potential component failures, ARI can proactively address these issues while the vehicle is being serviced for routine maintenance, thus minimizing catastrophic repairs and avoiding roadside trouble calls. As Fig. 4.4 illustrates, algorithms compute the likelihood of failure using historical maintenance data combined with vehicle sensor information supplied via telematics devices, odometer predictions, and driving patterns. In Fig. 4.3, the top three component expense categories (tires, engine, transmission, etc.) are categorized by manufacturer, cost, and occurrences to highlight potential patterns. The software allows the user to drill through the aggregate data into the details and to change the dimensions and measures as new questions arise.

To further maximize asset uptime, telematics data are now used to determine when a vehicle is repaired and back in service. Historically, the only method for learning whether a vendor was finished repairing an asset was to call them within a predetermined timeframe. Requiring vendors to notify ARI has proved to be a futile effort. ARI's vendors are geocoded, meaning their longitude and latitude are recorded in their system. When a vehicle initially goes to a vendor, a geofence is

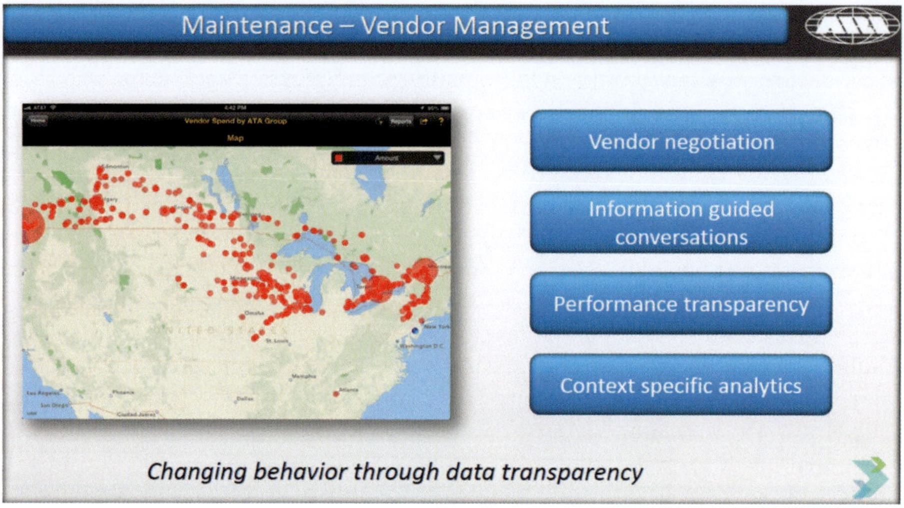

Fig. 4.4 Vendor management analytics

constructed by matching the asset's position to the vendor's location. Once the asset is repaired and the driver is back on the road, the geofence is broken, and ARI's system is automatically notified that the asset is back in service – completely negating the need for any phone calls.

Proper vendor management is necessary to ensure that ARI's customers receive the best service at the best price. ARI's vendor management team relies on vendor analytics to determine which vendors are performing well and which ones need some guidance and/or counseling. The team is armed with information to identify vendors with repair prices that frequently fall outside the norm and also details vendors who receive poor ratings from ARI's call center technicians, driver/customer feedback, and scores from the VMS system's vendor performance algorithms. The vendor management team is very transparent with the data and uses the information during conversations to show the vendors how they are performing and compares and contrasts them to other similar maintenance vendors within a specific proximity using geospatial capabilities. Vendors can easily ascertain if their prices are in line with their competition. If they are not and are unwilling to negotiate lower rates, ARI will typically steer future work to another vendor.

Vendor scorecards that measure performance, cost to value, customer service experience, and first-time resolution are key components of ARI's customer facing insight system. Fleet managers can quickly and easily determine which vendors are performing per expectations and which ones are the outliers. ARI frequently reviews vendor performance with each vendor to ensure they understand expectations and are meeting objectives. Figure 4.4 presents an example of a vendor analysis that identifies specific repairs by highest cost. This information is reviewed during vendor relationship meetings to ensure that the repair facilities provide a fair and competitive price relative to their geographical footprint.

Fig. 4.5 Key performance indicators in fuel management analytics

Aligning Fuel Management Strategy with Analytics

After the asset's monthly depreciation, fuel is the largest expense. Given that it is a variable cost, its volatility can lead to budgeting challenges. Fleets that utilize fuel cards benefit from tighter fuel controls via detailed transactional information. As Fig. 4.5 depicts, this information is analyzed for anomalies, fraud, and cost-saving opportunities. The key performance indicators in Fig. 4.6 can be customized based on each customer's objectives. As an example, this particular customer has chosen to be alerted if any vehicle purchases more fuel than it can store (tank violations). This alert can indicate possible pilferage. Another key performance indicator on the customer's dashboard identifies the percentage of assets with a company fuel card. If the asset lacks a fuel card, its transactions cannot be measured. Finally, the customer is also concerned with fuel economy and carbon dioxide outputs, which have an inverse relationship. As the fuel economy increases, the asset's carbon footprint decreases.

As an example, fleet managers typically look for assets that have had multiple fuel transactions in a single day, transactions where the gallons purchased exceed the asset's capacity, drivers purchasing unwarranted premium fuel, and transactions on a consecutive Friday and Monday, indicating the asset was used during the weekend (unauthorized use). Finally, as Fig. 4.6 demonstrates, by combining the data about daily fuel rates from suppliers with the fleet's geospatial footprint, drivers can be steered toward the lowest fuel provider or station based on their needs.

Additional auditing controls on assets with telematics devices can be brought to bear for both maintenance and fuel transactions. As an example, when a fuel transaction is sent to ARI, it is assumed the transaction was for the company vehicle. However, how would the company know if the driver decided to use the company fuel card to fuel his/her spouse's vehicle? If the vehicle is equipped with telematics,

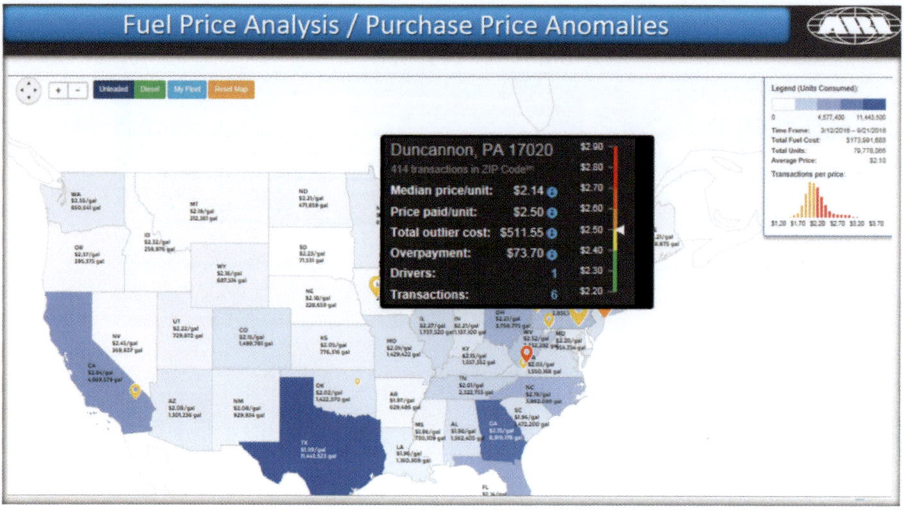

Fig. 4.6 Fuel price analytics management strategy

ARI can easily and accurately determine if the company vehicle was physically at the gas station when the transaction was authorized by comparing the asset's geolocation to that of the vendor. If a disparity is identified, the company can investigate whether a fraudulent transaction transpired to avoid future losses.

Aligning Operator Performance with Analytics

Determining how the driver is operating the asset is vital to cost control and its effective and safe use. ARI's driver safety program collates information from a variety of sources to understand the driver and his/her propensities. Motor vehicle records are analyzed in conjunction with maintenance and fuel use, driving patterns (hard stops, turns, speeding, etc.), and accident information to determine and rate the driver's performance. Based on the algorithms, specific driver training modules are assigned to educate and change behavior before an incident occurs.

ARI provides its customers' drivers with a mobile application that can be used to locate the most cost-effective fueling station and repair facility within a specified proximity based on type of fuel (unleaded, diesel, LPG, etc.) and type of repair required (oil change, alignment, major component failure, etc.), and identifies the preferred vendor(s) based on key performance indicators related to quality of work, customer service experience, and cost to value. The application is a pull method, meaning, the drivers need to log in to the mobile application to request information. This action was chosen instead of a push/text method to avoid the possibility of distracted driving. Thus, aligning operator operations with analytics promotes the cost-effectiveness of fleet operations.

Use Case Example: Red Hawk Fire & Security LLC

Red Hawk Fire & Security LLC is an existing ARI customer that operates a fleet in excess of 800 trucks in the USA in support of their fire and security products. In 2013 their expenses from preventable accidents were a staggering $1.5 million. ARI was consulted (SiliconAngle, 2017) and, working together with Red Hawk, devised a multistep process to first understand the contributing factors and then implement technology to monitor, manage, and reduce their accident expense. Telematics devices were installed on their fleet. By leveraging their information and combining it with motor vehicle reports, ARI was able to identify high-risk drivers, who were required to take specific driver training courses to change their behavior. As a result, Red Hawk's accident expenses were reduced from $1.5 million in 2013 to $350 K in 2015 and as of 2016 dropped to $250 K.

Future Trends: Advanced Analytics Technology at ARI

ARI invests heavily in its analytic capabilities, investing about 25 percent of its operating budget in systems and technology. The fleet management market is unique. Off-the-shelf software packages are typically not available, so systems need to be custom designed to provide the capabilities required to tackle difficult business problems while providing a competitive advantage. Five years ago, ARI began investing in a new state-of-the-art analytic foundation in order to capitalize on its strategy. Major capital investments were made across the board with significant acquisitions of SAP, Oracle, and EMC technology. As Fig. 4.7 demonstrates, SAP HANA, an in-memory database platform, was chosen as the analytic foundation. HANA's remarkable speed was one of the main decision points. As data volumes increased and became more verbose, ARI wanted to ensure that the user experience would not be affected. Reports and analyses that used to take minutes or hours, and in some cases never returned a result, now consistently run in milliseconds. Speed is a powerful catalyst for innovation – it removes barriers and enables creativity.

Moving beyond the foundational aspects, ARI wanted to ensure that data analytics was pervasive throughout the organization – its full potential would not be realized if it remained an IT capability. To that end, it utilized SAP Web Intelligence for reporting and analysis and Lumira for visual exploration. To further leverage the investment, ARI's. NET development group has begun creating new applications directly on HANA. Using this multipronged approach, a culture of data-driven decisions is leading to innovative new products and services and improved customer service.

ARI's analytic roadmap is full of challenges and opportunities and is centered on extending its HANA investment. Sentiment analysis, enhanced statistical capabilities with "R" and HANA's PAL (predictive analytic library), and embedded geospa-

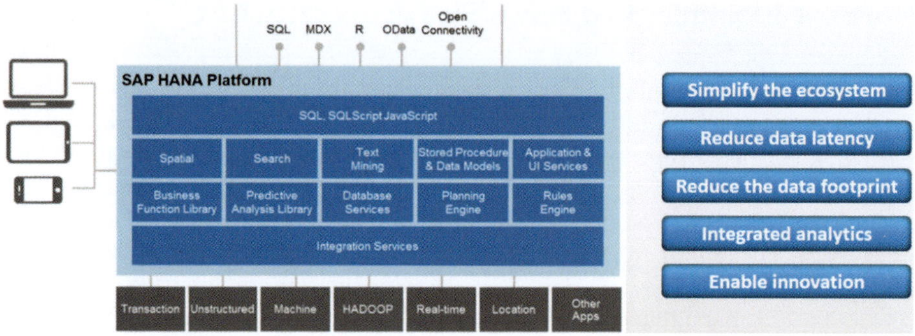

Fig. 4.7 Advanced technology analytics overview at ARI

tial capabilities are but a few of the items that the team will be exploiting over the next 12 months.

In addition to the enhanced analytic capabilities, ARI is also exploring the convergence of its transactional and analytic platforms into HANA. Previously, system architecture's best practices consisted of separating transactional and reporting and analytic databases. However, HANA's unique design now allows these disparate systems to be combined into a single platform. The result is a simplified architecture that removes data latency (the time it takes to transfer data from the transactional to the reporting database) and reduces the overall data footprint, thereby lowering the operating costs.

SAP is a strategic vendor of ARI providing insightful guidance in terms of technology market trends and helping to establish relationships with other customers in the automotive space where best practices are discussed and shared. ARI frequently provides key input at SAP events and conferences (Powell, 2017), detailing the alignment of technology and strategy to solve complex business challenges. These opportunities provide a conduit to a larger market using SAP's connections. They produce positive results for SAP and ARI.

Bridging the Gap Between Theory and Practice

Addressing the challenges and opportunities inherent in fleet management can best be accomplished by creating a forum or round table of industry representatives and academics to create an awareness of the issues at hand. Both traditional and nontraditional solutions provided by academia would be invaluable to the fleet management community. Fleet management is at an inflection point where margins are being challenged, and differentiation between competitors is becoming blurred. Having an outside view and incorporating nontraditional perspectives into the fleet business is a healthy exercise and should be leveraged to provide opportunities where tunnel vision has set in. Leading companies in the fleet management industry such as ARI could take the lead in organizing and hosting such a forum for educators to give them an inside look into the business to determine where opportunities exist and efforts are best placed. The key is to build advocates within the business community. Once the value is recognized, the requests for better alignment with academia will grow organically.

Another area that needs to be addressed is how best to prepare students to address current business problems in the fleet management industry. We believe that students can be better prepared to handle complex business problems through hands-on, real-life exposure. Internship programs such as Drexel University's co-op program are an excellent method to expose students to the challenges facing local businesses and give them experiences beyond the classroom environment. Another suggestion is embedding elements of the fleet management business into the students' curriculum through programs such as business analytics. On-campus presentations are also another excellent method to provide awareness about fleet management to students. These presentations can be broad based or more specific to the business and the students' area of interest (supply chain, analytics, marketing, etc.). Finally, offering business problem-solving and consulting courses to students in which fleet companies provide real data and seek solutions is another proven way to prepare students for the business.

Steps to Implement a Telematics and Analytics Strategy

Implementing an analytics strategy, whether for telematics or any other type of product or service, starts with a business need. Businesses should begin with actionable questions that need to be investigated to improve efficiencies, increase customer retention rates, reduce costs, identify and create new revenue streams, and enhance products and services. Strategies that begin with technology and then try to find business problems to solve are destined to fail. All successful projects need advocates, and analytic projects are no exception. Advocates will help sell the value and demonstrate the impact to other business leaders in contexts and forums to which IT might not have access. The key is to have the business drive the need and sell the value to build momentum. If information technology is pushing the project and/or if IT is working in a silo, the chances for failure are exponentially raised.

Once the advocate(s) and business problems and questions have been identified and agreed to, a clear and measurable understanding of success should be established and shared with the team. Initial steps should be small. Large challenges should not be confronted until incremental success stories are created. Such stories will be needed to counter resistance.

At this stage, attention should turn inward to determine the underlying data requirements, how the information will be shared (reports, visual analysis, integrated into existing transactional systems, etc.), the data's quality, the necessary human resources with the appropriate skillset to ensure success (architects, statisticians, data modelers, etc.), and the proper controls to implement related to data privacy and security and its ecosystem (creation, use, storage, and destruction). It is critical to review these items in depth and to look at their costs and dependencies from a holistic perspective. Otherwise, tactical solutions will have a tendency to emerge instead of a strategic approach.

Conclusion: Enabling Effective Fleet Management Strategies with Analytics

As described in this paper, advanced analytics (TDWI, 2014) plays an important role in fleet management solutions and makes numerous insights possible (Network Fleet, 2015):

1. Predictive analytics – Usage of historic and current information to calculate the likelihood and probability of events such as speeding tickets.
2. Prescriptive analytics – Details specific output based on inputs. An example would be a driver receiving a warning that he or she was about to go beyond a set number of speeding violations.
3. Geospatial analytics – Usage of spatial data to analyze events such as fleet location tracking.
4. Operational intelligence – Enhancing fleet operations by continuously analyzing real-time data leading to operational efficiency.

Like other types of risk assessment programs, the usage of analytics to derive insights requires commitment from all parts of the organization. As Fig. 4.8 illustrates, the most successful users are organizations that involve all departments including operations, finance, marketing, safety, HR, and IT in the process at an early stage. One reason previous predictive analytics projects have failed in the past is due to lack of alignment and enterprise-wide adoption (Automotive Fleet, 2015a).

In order for analytics to reach its full potential and become pervasive within an organization, it needs to have the appropriate presence so that no one line-of-business steers its direction. Our experience reflects an analytics discipline that typically begins in IT but quickly grows beyond its borders once advocates from other lines-of-business surface and early achievements demonstrate its value. Experience has shown that a federated organizational approach involving building relationships with key stakeholders is more effective than a centralized service where requests for analyses are performed by one group, typically in IT. A federated approach (hub and spoke) has the added advantage of empowering all departments, highlighting best practices and techniques, and removing the typical funnel effect associated with centralized services that can greatly impede the business's ability to scale.

Identifying and hiring the right analytics talent is by far the most challenging aspect of instilling an analytic discipline within an organization. ARI has taken a two-pronged approach to hiring and developing talent in this space. First, current employees who have the right mix of desire, aptitude, business acumen, and IT

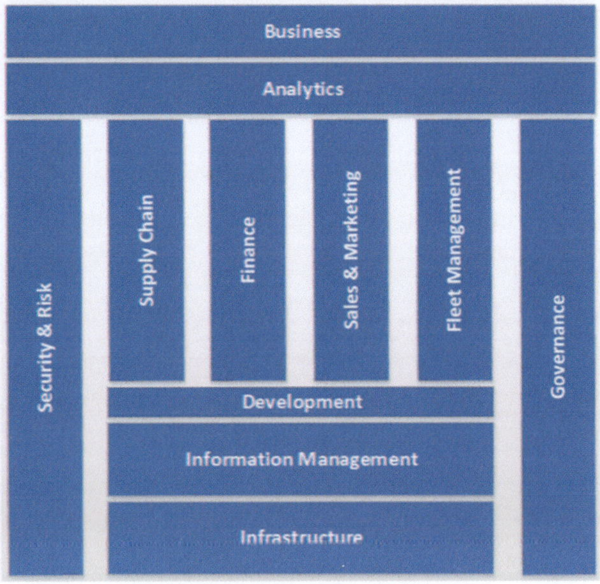

Fig. 4.8 Organizational alignment framework and role of analytics

capability are actively recruited. Using a combination of in-house educational services and third-party training, these employees' skills are enhanced to maximize their performance. The second path ARI has taken is to align itself with universities in the Philadelphia tristate area. Drexel University was chosen as the primary school due to their well-respected analytics program, their proximity to ARI, Drexel's co-op program, and overall support from multiple Drexel alumni at ARI.

Top Mistakes to Avoid When Gathering Data

Here are some mistakes to avoid in data gathering (Automotive Fleet, 2015b) for fleet analytics:

- Not differentiating between fleet vehicles – Fleet vehicles are not uniform. They have their own requirements that need to be considered.
- Focusing on the wrong issues – The business proposition has to be proven before using data and analytics.
- Using an incorrect sample size – Ensuring that the sample size is large enough helps avoid problems.
- Using questionable data – The importance of checks on the data quality is critical to model development.

Top Mistakes to Avoid When Analyzing Data

- Becoming overwhelmed by voluminous data
- Using incomplete data in analysis
- Focusing too much on numbers without understanding their origin
- Not paying attention to outliers and anomalies
- Ignoring industry benchmarks in performance
- Not understanding all of the variables and drawing incorrect insights and conclusions
- Not considering enterprise-wide solutions and opportunities

Understanding the business challenge and value proposition and then collecting data to input into models that solve a real-world problem is critical to success.

References

Automotive Fleet. (2015a, July). How to leverage constantly evolving fleet analytics. *Automotive Fleet*. Retrieved from http://www.automotive-fleet.com/article/story/2015/07/how-to-leverage-constantly-evolving-fleet-analytics.aspx

Automotive Fleet. (2015b, September). Using Predictive Analytics to improve fleet decisions. *Automotive Fleet*. Retrieved from http://www.automotive-fleet.com/channel/utility/article/story/2015/10/using-predictive-analytics.aspx

Network Fleet. (2015). Advanced analytics in fleet management: The how and why. *Network Fleet White Paper*. Retrieved from http://info.networkfleet.com/rs/785-DCW-685/images/WP-N037%20Analytics.pdf?aliId=62493019

Powell. (2017). *SAP Leonardo Live Conference*. Retrieved from http://www.arifleet.com/all-news/according-to-forbes-sap-leonardo-live-featuring-aris-bill-powell-is-a-top-conference-not-to-be-missed/; https://www.forbes.com/sites/stevebanker/2017/07/25/put-leonardo-live-on-the-list-of-key-business-conferences-to-attend/#17a68f105627

PWC. (2015). PWC's fleet management solution. *PWC White Paper*. Retrieved from https://www.pwc.com/us/en/increasing-it-effectiveness/assets/pwc-ms-fleet-management-solution.pdf

SiliconAngle. (2017). How the internet of things slashed one fleet owner's accident cost. SiliconAngle Article. Retrieved from http://siliconangle.com/blog/2017/01/02/internet-things-slashed-fleet-owners-accident-costs-85-percent/

TDWI. (2014, December 18). Best practices report: Next-generation analytics and platforms. *TDWI*. Retrieved from https://tdwi.org/research/2014/12/best-practices-report-next-generation-analytics-and-platforms.aspx

Chapter 5
Aligning Data Analytics and Supply Chain Strategy in the Biopharmaceutical Industry

Mark Holder, Amit Devpura, Anthony Lee, and Suresh Chandran

Abstract Much has been written recently about the important role that data and analytics will play in improving productivity and profitability of companies in the biopharmaceutical industry. Data analytics will be a source for value creation and sustained competitive advantage for companies as new technologies like the Internet of Things and digitization of supply chain play a role in transitioning this industry into a more customer-centric model. This paper provides an overview of the status of the pharmaceutical industry and role that data analytics plays in supply chain management. The objective of this paper is to provide a use case example of implementation of a supply chain blueprint model including specifics of technology platforms, planning and optimization tools, and value stream mapping that have enabled tremendous cost savings at AstraZeneca. Lessons learned from experience with consulting to other companies in the biopharmaceutical space in the area of data analytics and strategy are outlined. The importance of fostering a two-way dialogue between members of the business community and educators and introducing new programs like the future leaders program and Supply Chain Boards in bridging the gap between theory and practice through meaningful partnerships is also discussed.

Keywords Pharmaceutical industry · Supply chain management · Data analytics · Technology platforms · Planning and optimization tools · Value stream mapping

---

M. Holder · A. Devpura
AstraZeneca, Cambridge, UK
e-mail: mark.holder@astrazeneca.com; Amit.devpura@astrazeneca.com

A. Lee
A4P Inc., Philadelphia, PA, USA

S. Chandran (✉)
Drexel University, Philadelphia, PA, USA
e-mail: sc44@drexel.edu

M. Anandarajan, T. D. Harrison (eds.), *Aligning Business Strategies and Analytics*, Advances in Analytics and Data Science 1,
https://doi.org/10.1007/978-3-319-93299-6_5

Introduction

Big data is unique because of the volume generated, variety of data types (including texts, pictures, videos), and the high velocity of data processing, which today is both widely available and much less expensive to access and store (McAfee & Brynjolfsson, 2012). Big data is more than a buzzword; it is a competitive resource with significant implications to company strategy and alignment with the stakeholder-based view of the firm. Big data are being used to transform industries with better decision-making capabilities that allow for improved profitability (Mayer-Schönberger & Cukier, 2013). Based on their large-scale study, McAfee and Brynjolfsson (2012) note that companies which characterized themselves as data-driven were more productive and more profitable than their competitors. In the pharmaceutical industry, applications of big data are transforming all aspects from drug discovery to development to marketing (Copping & Li, 2016).

Big data have the potential to revolutionize supply chain dynamics. To make the most of the big data revolution, supply chain leaders need to understand and embrace data science and implications for supply chain decision-making (Waller & Fawcett, 2013). Specifically from a supply chain strategy perspective, data science can be applied in qualitative and quantitative ways to solve problems in supply chain as well as be used as a predictive tool in strategically improving company performance and providing sustained competitive advantage. The objective of this paper is to seek to understand how best to align supply chain strategy and data analytics for value creation in the biopharmaceutical industry. A review of the present status of the industry and the important role that data analytics is expected to play in supply chain strategy is outlined. The experience with consulting to companies in the pharmaceutical industry and lessons learned in linking data analytics to supply chain strategy is reviewed. The evolution and development of the supply chain blueprint and importance of data analytics are detailed in a case study of AstraZeneca, a global biopharmaceutical company. The role that people, process, and technology play to enable operational cost savings in supply chain at AstraZeneca is reviewed. We also look at how best to bridge the gap between theory and practice as we look at emerging trends in linking supply chain strategy to data analytics.

Background

AstraZeneca is a global, science-led biopharmaceutical company operating in a variety of therapeutic areas (oncology, respiratory, neuroscience, cardiovascular, etc.) and developing medicines that are used by millions of patients worldwide (AstraZeneca, 2017).

Biopharmaceutical companies produce medicines that have biological basis. Thus biopharmaceutical companies use living organisms like bacteria, yeast, etc. to manufacture medicines as opposed to chemical synthesis that pharmaceutical companies use.

Biopharmaceutical companies are a rapidly growing subset of the pharmaceutical industry and make up about 20% of the pharmaceutical market (Otto, Santagostino, & Schrader, 2014). The $967 billion global pharmaceutical industry (Statista, 2017) has been experiencing an increase in the volume, variety, and velocity of information as every aspect of the value chain in this industry has undergone examination for cost and efficiency even as patient expectations are changing. This industry which has been struggling with growth and operational efficiency is transitioning to a more consumer-centric industry. In the USA, pharmaceutical companies, insurance companies, and hospitals are in the midst of consolidation to enable value (rather than volume)-based outcomes that call for lowering prices, streamlining for operational efficiency and convenience. This industry is increasingly recognizing the importance of linking data and analytics to business strategy for value creation (Mentesana, Rotz, Strang, & Swanick, 2017).

The emergence of business analytics as a critical business tool for value creation and competitive advantage is due to the widespread use of digital technologies (Chen, Chiang, & Storey, 2012). Digital is transforming the world of business, and the pharmaceutical industry is no exception. While big data deals with the ability to process data that has velocity, variety, and volume, business analytics refers to using mathematical, statistical, and optimization techniques for deriving insight (Accenture, 2014) to enable organizations make better business decisions (Muhtaroglu, Demir, Obali, & Girgin, 2013).

Supply chain for the pharmaceutical industry involves all value-added activities needed to manufacture the product to be delivered to the marketplace (whether to the hospital, pharmacy, or patient). Traditionally the supply chain of pharmaceutical companies has been complex and not very cost-effective. In the era of blockbuster drugs, the supply chain problems were not as serious, but in today's competitive environment with patent expirations denting revenues and move toward outcome-based thresholds, this area needs a serious relook. There are numerous forces reshaping the pharmaceutical environment: diversity in products and processing technologies, shorter product lifecycles, ability to scale up and down rapidly, growing importance of emerging markets, environmental regulations, greater need for risk management, etc. which calls for radically changing supply chain operations (PWC, 2017).

While traditional supply chain planning systems have focused on day-to-day operations and have limited analytical (modeling and forecasting) capabilities, advanced analytical systems using mathematical, statistical, optimization, and visualization tools have predictive capabilities thereby allowing for real-time insights and a proactive response to optimize supply chain decisions (SAS 2010). The pharmaceutical industry is aggressively looking to implement new technologies and concepts like Internet of Things (IoT), digital supply chain, and Supply Chain 4.0, particularly in logistics and supply chain management (Wamba, Akter, Edwards, Chopin, & Gnanzou, 2015), to improve operational efficiencies while meeting patient needs better. The value of analytics in logistics and supply chain operations is to improve the visibility, flexibility, and integration of global supply chains in terms of sourcing and logistics (Genpact, 2014), inventory management, handling cost fluctuations, and demand planning (Wang, Gunasekaran, Ngai, & Papadopoulos, 2016).

Aligning Business Strategy and Analytics

There are three key advancements in the use of big data that have helped industries capture value using data and analytics:

- The invention, proliferating adoption, and ease of use of the Internet
- Continued rapid cost reduction of computing processing and storage products
- Exponential technologies and computing power improvements

These advancements have led to an explosion in data generation and in a lot of cases an analytics gap as relates to strategy. The issues related to aligning strategy and analytics depend not only on business process and integration but also on where each business is along the spectrum in the adoption and application of information technology in the big data world today.

Lessons Learned from Consulting in the Pharmaceutical Industry

Some examples of lessons learned (Lee, 2016) from consulting to companies in the pharmaceutical industry on linking analytics and strategy are outlined below:

1. Importance of developing realistic trends – This was proved for a global pharmaceutical company's subsidiary operating in a developing country where regulatory and government policy changes derailed the old methodology affecting budgets and projections. Analytics helped quickly reset issues in the supply chain area by developing realistic projections in the new (changed) environment.
2. Improving forecast accuracies in the supply chain – Analytics was used to improve forecasting accuracies by 30–50% in a US-based pharmaceutical company's supply chain. Previous forecasts were hampered by a lack of focus and scope in aligning business strategy and analytics because individual departments viewed certain key metrics as irrelevant to them. The solution implemented called for commercial strategies to be implemented with monthly regional demand planning through integrated processes, analytical tools, and cross-department accountabilities. Having key business leaders actively involved in the process and the importance of organizational structure and communication were highlighted during this experience.

Value of a holistic approach to implementation of a fact-based analytics strategy – This was highlighted for a US-based pharmaceutical company during implementation of best practice solutions that used analytics to meet strategy, goals, and objectives. The importance of cross-functional interactions (rather than working in isolated silos) and communication to implement planning processes for value creation was evident during this exercise. As shown in Fig. 5.1, forecasting errors over a 10-year period for this company were reduced to 2% (from 20% previously), and

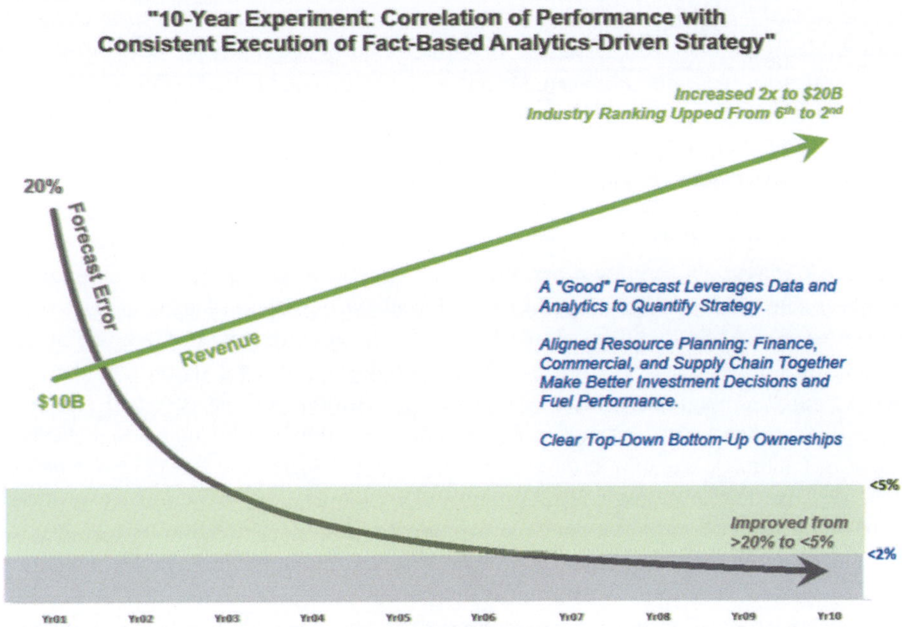

Fig. 5.1 Result of execution of analytics-based strategy

the company climbed to 2nd place (from 6th place) in industry rankings by implementation and execution of a fact-based analytics-driven strategy. Getting various functional areas like finance, commercial, and supply chain and operations to work together and align with company strategy was key to progress.

Previously, it was not uncommon for pharmaceutical companies to custom design and build in house systems so that the right kind of data could be captured to develop insightful analytics and *improve forecast accuracies* so as to inform product management teams of opportunities and challenges. Today's software offerings address those gaps and add more capabilities. What was limited to, say, a single-country and offline application in the past can now be implemented across multi-countries enterprise and in real-time online. In addition, technologies such as business intelligence tools are also helping to close the analytics gap. These tools bypass the IT chasm and shorten the "distance" between the decision-makers and data while being able to surface information and analytics in much shorter time or in real time.

The key to success in practical implementation of data and analytics calls for using the right tools. During one implementation (Lee, 2016), Excel was selected to perform and integrate the analytics across a handful of entities. Conceptually, the logic and integration in the monthly process were straightforward. However, operationally, the time pressure and volume of disparate data involved became very cumbersome and eventually overwhelmed the analyst and led to resistance for a follow-up improvement project. Having the right tools that can easily and quickly accomplish otherwise cumbersome and mundane work can garner more support for successful practical implementation. Software vendors also seem to have evolved into two camps – ones that start from scratch and develop advanced data mining capabilities in the software's foundation versus more established software vendors that still maintain traditional data management technologies but dress it up with a flexible user interface. The former group is more agile and holds promise in capabilities to extract newer insights.

Another area that is key to success in practical implementation of data analytics relates to gaining senior level support (Lee, 2016). While bottom-up on-the-ground support is important, top-down organizational support is key to success. Ultimately, people make the final planning decisions based on perspectives gained from their work experience. And while analytics could help to minimize the influence of biases in decision-making, real adherence to that desired practice depends on how directly and strongly senior management articulates and enforces the process and practice. The 10-year experiment, outlined in Fig. 5.1, occurred during a period when that organization's senior leadership strongly believed and articulated the use of a fact-based analytics practice for strategy development and implementation. Experience elsewhere where there is weaker senior management interest and/or support in the analytics-based decision-making practice seems to correlate with less stellar business performance.

This experience from consulting in the pharmaceutical industry and those outlined in surveys (Accenture, 2014) suggest that overall success comes from having an enterprise-wide strategy that uses data to drive business value. Embedding analytics in the day-to-day supply chain and operationalizing it rather than using analytics on an ad hoc basis are key to operational success. Another interesting finding is that as the three Vs of data – volume, variety, and velocity – keep increasing along with the complexity in supply chain, there is a feeling among supply chain leaders that their investment in analytics is not up to the mark and more investment in analytics is needed (Gartner, 2017). The reality, however, is that 50% of existing supply chain planning solutions in these organizations are not being utilized because these companies have not reached the higher levels of maturity required for more advanced analytics to be deployed (Gartner, 2017). These organizations need very basic levels of foundational capabilities like reports, dashboards, scenario, and statistical analysis tools in their toolbox for implementing requisite solutions. It is very important that these organizations look at the problem to be solved and business value that can be created by deploying supply chain analytics whether for managing production, operations, inventory, demand, or forecasting.

Case Study of Aligning Supply Chain Strategy with Data Analytics: AstraZeneca

AstraZeneca is a global, science-led biopharmaceutical company that develops innovative life-changing medicines. The scientific and data-driven approach is in full display across different functions across the company including supply chain. The evolution of supply chain (SC) and the use of analytics at AstraZeneca have

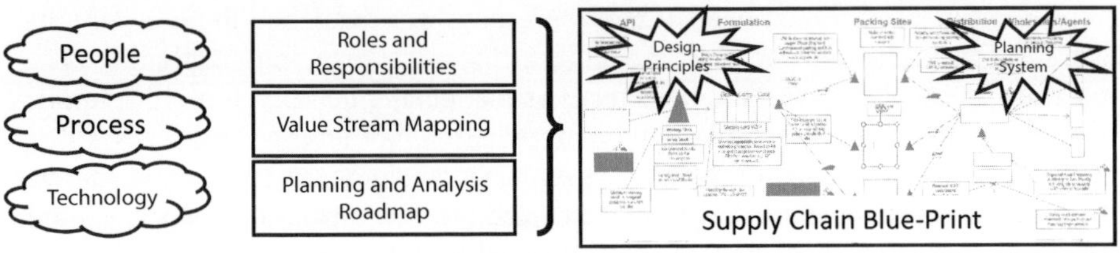

Fig. 5.2 Supply chain blueprint at AstraZeneca

been a journey. Significant progress has been made over the years, by thoughtfully evaluating the level of maturity and then implementing the right analytical technique to best support decision-making. Starting with simple and crude brute-force methods, the company has matured into using sophisticated data science for SC.

Data and analytics is used in various aspects of supply chain management, of which we are going to highlight a few vis-à-vis, developing the supply chain blueprint (SCBP), the way the company does it long range capacity planning, determines the right inventory levels, and its quest to improve velocity while decreasing variability.

The process of utilizing and developing a SCBP has to be underpinned by three key areas, people, process, and technology, as shown in Fig. 5.2. The people approach is an organizational structure and focuses on supply chain excellence matched with the capabilities to execute the SCBP strategies. Process means globally aligned and governed processes that are standardized and followed. Technology works as an enabler and needs to be a technology platform sufficiently developed and robust in capability that enables and supports the access of critical business data and analytics needed for supply chain excellence. On the people and process side, new roles were created; responsibilities, streamlines, and interactions formalized; and expectations clarified. On the technical side, technology platforms in modern supply chains serve an important enabling function. Without an end to end enterprise resource planning system implemented in conjunction with standardized process, most supply chains would be very difficult to coordinate and execute. These systems are enablers of key supply chain processes and allow resources and assets to be efficiently utilized while alerting supply chain planners to potential problems with their plans or schedules. Over the past 3 years, AstraZeneca has made a very large investment in upgrading and aligning all of their SAP implementations within AstraZeneca to be on a common platform to drive standardization of their supply planning and demand planning processes and provide system-wide visibility to inventory, capacity outlook, and key planning parameters. This upgrade also includes a global roll out of an APO (advanced planning and optimization) tool. Without this investment and the visibility it now provides, achieving much lasting change would be very challenging. This technology upgrade in conjunction with the value stream mapping (VSM), a LEAN process, and implementation of the SCBP has enabled tremendous savings in their supply chains. The key outcomes measured are inventory level reductions, improved total lead time, reduced stockouts, and improved risk management, reporting to show the benefits.

As the blueprint got embedded and more data was available, the company has over time started using statistical modeling for service level optimization by effective trade-off between cost and inventory. As a company that develops and supplies innovative medicines, the availability is extremely important as in most cases there is no alternative available to patients. To keep the medicines cost-effective to patients, the company uses a scientific approach to determine how much safety stock it should carry. Using product history, life cycle stage of the medicine, demand and supply variability, and desired customer service levels, the right inventory target is determined using a blend of statistical techniques. By being able to group products effectively, quantitative determination of cycle time and cycle time variability, analytically determined replenishment intervals, and use of safety formula, the company has achieved one of the best inventory performances in the industry. Today the company uses a custom tool for inventory optimization which has been developed as a bolt-on to SAP-APO, with ongoing work to implement MEIO capability using Llamasoft.

As the data maturity grew, the company initiated the use of data science in strategic supply chain decisions. In the highly regulated pharmaceutical industry, it is vital to operate as efficiently as possible to maximize the availability of high-quality medicines to patients at minimal cost. This starts with having the right manufacturing capacity and plans for the future. To support the changing product portfolio of the company, capital decisions needed to be made quickly while analyzing multiple options and scenarios. So, to help determine the long-term operations investment strategy, a blend of techniques using Monte Carlo simulation and an integer programming-based solution is being used. A dedicated team of experts evaluated different options and settled on Crystal Ball for Monte Carlo simulations and Llamasoft for capacity planning and optimization. They then collected detailed manufacturing data from the sites, which along with the 10-year demand forecast is used to develop and validate the base model. Then different scenarios were analyzed to evaluate potential impact of successes of different therapies, changes to demand volume and geographic distribution, impacts of divestitures and acquisitions, and impacts of expansion, consolidation, and rationalization of capacity. The rigorous analysis provides a reliable and consistent approach to future capacity planning. The model has been used to evaluate and recommend several network changes to best meet the long-term needs of the company and patients. The solution has not only highlighted potential financial benefit but has been instrumental in instilling a more conscientious data-driven approach to decision-making.

Another place where data science is seen in use at AstraZeneca is in how the company is improving the velocity with which products flow through the supply chain while reducing the variability in the process. The whole process of getting to the right information is a momentous task as there are large volumes of data distributed across multiple systems across the globe. With the distributed nature of the manufacturing process, a product might start at one site and go through different processing at multiple other sites. Integrating and deriving conclusions from this data that resides at these different locations with its own data specifications and nomenclature are a big data exercise. The information from different ERP systems is first accumulated in a data lake, and then different algorithms run to create relationship between the large volumes of manufacturing, shipping, and sales data, thus determining the true velocity and variability in the process. Most of the data is processed through structured database built in Oracle and is visualized through QlikView. Having the standard way to process the data and then display the output is critical to diagnose the issues and make the results actionable. This information is then used to compare performance, identify opportunities, and track progress for the various velocity improvement plans. With these activities, the company has already reduced WIP and inventory by $200 million and expects further reduction of more than $500 million over the next 3 years.

Emerging and Future Trends

There are enormous opportunities available for pharmaceutical companies to link data and analytics to business value creation and performance improvement. A few examples opportunities are outlined (Champagne, Hung, & Leclerc, 2015) below:

- Research and development will involve more modeling and computer simulation techniques, to enhance organizational effectiveness.
- Marketing and sales will work to better understand customers including cures for rare diseases.
- More linkages of data within players in the pharmaceutical ecosystem will become commonplace as more digital tools will be used to evaluate best treatment options to meet outcome-based thresholds.

Bridging the Gap Between Theory and Practice

One way to bridge the gap between theory and practice is the use of future leaders programs. The opportunity to bring on new young talent and provide a supportive environment to allow these new recruits to evolve into future leaders is a challenge many pharmaceutical companies are facing. A creative and successful way some companies are dealing with this challenge is the use of a targeted program that enrolls recent college graduates into a controlled and focused future leaders development program. The first step in the success of these future leaders program is to bring in high-caliber graduates, with potential and ambition to become a future leader at the company. The use of a rigorous screening and hiring process that includes a behavior-based interview, case study, real-life problem-solving, and group interactions with other possible recruits ensures the pool of new hires is well qualified to meet the rigors of the 3-year future leaders program. Secondly, by ensuring each individual that enters the program has a home organizational sponsor and a senior level mentor, recruits' success rates are increased greatly. Additionally, the home organization "owns the head count" at the end of the 3-year program thus guaranteeing the recruit a role upon completion of the program if they have not landed a role or been exited from the program. By including this requirement, it allows the recruit to focus on embracing the learning and developing experience and for delivering value in the roles they rotate through as well as ensures the home organization stays engaged with the recruit progress and success throughout the program.

In one pharmaceutical company's program, a key element of their future leaders program is leadership development. In the first 2 years, a recruit will be split between three different placements, and each role will give them valuable insight on how to influence decisions and provide an opportunity to learn how that area of the business operates and connects with other parts of the business. Each recruit will also spend a minimum of 3 months living and working abroad, to gain exposure to different cultures and ways of working, as a hands-on way to test and improve their ability to handle responsibility. Finally, the recruit will land a consolidation role that will be their final role in the program and becomes their first role as a full-time employee. By giving recruits a vital pharmaceutical industry exposure, this future leaders development program is the perfect springboard to a senior job opportunity within the company.

The future leaders development program approach provides a number of benefits to both the company and the recruits. First, and most importantly, this approach allows the new recruits to effectively transition into the corporate world from the academic environment in a supportive and controlled way. By placing these recruits into meaningful roles and rotating them through assignments over a 3-year period, the recruit gets a chance to better understand the career path they prefer to follow that is aligned with their aspirations and strengths. The company gets an extended period to determine which area of the organization the new recruit will have the best chance for future success or if they should be exited from the program. This much longer and more in-depth review and evaluation of a recruit's future potential ensures the best candidates are identified for retention and assigned the more challenging roles in areas that match their aspirations. This extended "grace" or "provisional" period helps both the recruit and the company ensure future success as a full-time employee and leader. Another organizational alignment benefit is the ability to match new young talent to areas of greatest organizational need due to organizational growth, introduction of a new business operating model, or shifts in business approach due to external business environment changes. This future leaders program approach also provides the new recruits early access to senior leaders in the company which promotes the accelerated development of leadership prospective and guidance on the path they may want to choose in their careers. Finally, the recruits benefit from the establishment a solid network with their counterparts in the program as well across many parts and levels of the organization with all the managers and employees they have worked with the 3-year program. This approach has promoted a much higher level of successful program completion and retention of future leaders and talent than other avenues or approaches.

Another way of bridging the gap between theory and practice is the establishment of Supply Chain Boards operated and managed by academic institutions and industry partners. By establishing these partnerships through a coordinated and structured approach that promote a shared responsibility for analyzing and studying industry practices and challenges, it allows for the application of academic theory in support resolution to these challenges and improvement of industry practices. Additionally, the development of this synergistic relationship between academia and industry ensures the continuous access to critical feedback on how the theory and practices being taught apply to real-world challenges and practices. Additionally, it allows for ongoing collaboration and feedback for driving improvement through meaningful academia and industry programs, symposiums, and training events. This ensures the problems supply chain organizations are dealing with in the industry have a partner in academia that is looking at the best way to solve it. These Supply Chain Board partnerships also extend to undergraduate students, graduate students, and professors in very beneficial ways. It allows access to industry resources to funnel research questions and queries to as well as gather real-world feedback on new approaches and theories. It provides outlets for student project teams to study real-world supply chain challenges. It promotes a strong partnership to open up intern and future employment opportunities for graduating students. Bridging the gap between theory and practice is done best when both industry and academia partner in meaningful ways to make this happen.

Conclusion

As the pharmaceutical industry supply chain continues to evolve in its application of data and analytics, due consideration must be given to the strategy and culture of the organization. Defining the business case is vital, and understanding that analytics needs to be linked with business strategy is critical. As outlined in the case with AstraZeneca, successful pharmaceutical companies will have organizational structures and processes in place that make them agile and ready for the future. Winners will not be defined not by who are the strongest today but by those that are quickest to adapt (Fox, Paley, Prevost, & Subramanian, 2016).

The big data world we live in today has the tools and technology to speed access to voluminous data and rush them into analytic engines to turn into information and insights to business people involved in developing strategies. The opportunities are plenty, but caution is also prudent. In the past, "More Is Better" was a default business motto when generating analytical reports. However, to succeed in the business world today, a different motto, "Less Is More," is needed (Lee, 2014). Getting there involves weeding out irrelevant data from voluminous data sets, and this can be achieved only if we align business strategy with the right analytics.

References

Accenture Global Operations Megatrends Study. (2014). *Big data analytics in supply chain: Hype or here to stay?* Retrieved from https://www.accenture.com/t20160106T194441__w__/fi-en/_acnmedia/Accenture/Conversion-Assets/DotCom/Documents/Global/PDF/Digital_1/Accenture-Global-Operations-Megatrends-Study-Big-Data-Analytics-v2.pdf

AstraZeneca. (2017). AstraZeneca Website. Retrieved from https://www.astrazeneca.com/our-company.html

Champagne, D., Hung, A., & Leclerc, O. (2015). The road to digital success in pharma. *McKinsey White Paper*. Retrieved from http://www.mckinsey.com/industries/pharmaceuticals-and-medical-products/our-insights/the-road-to-digital-success-in-pharma

Chen, H., Chiang, R., & Storey, V. (2012). Business intelligence and analytics: From big data to big impact. *MIS Quarterly, 4*(36), 1165–1188.

Copping, R., & Li, M. (2016). Analytics: The promise and challenge of big data for pharma. *Harvard Business Review*. Retrieved from https://hbr.org/2016/11/the-promise-and-challenge-of-big-data-for-pharma

Fox, B., Paley, A., Prevost, M., & Subramanian, N. (2016). Closing the digital gap in pharma. *McKinsey White Paper*. Retrieved from http://www.mckinsey.com/industries/pharmaceuticals-and-medical-products/our-insights/closing-the-digital-gap-in-pharma

Gartner. (2017). *When – and when not – to buy additional analytical capabilities for your supply chain planning process*. Retrieved from https://www.linkedin.com/pulse/when-buy-additional-analytical-capabilities-your-supply-pradhan

Genpact. (2014). Supply chain analytics. *Genpact White Paper*. Retrieved from http://www.genpact.com/docs/resource-/supply-chain-analytics

Lee, A. (2014). *Big data: An inevitable paradigm shift and two impacts for pharma*. Philadelphia, PA: Digital Pharma East Summit.

Lee, A. (2016). Accuracy for profitability – Is this achievable? *CBO Bio/Pharma Product Forecasting Summit*, Philadelphia, PA.

Mayer-Schönberger, V., & Cukier, K. (2013). *Big data: A revolution that will transform how we live, work, and think*. New York: Houghton Mifflin Harcourt Publishing Company.

McAfee, A., & Brynjolfsson, E. (2012). Big data: The management revolution. *Harvard Business Review, 90*, 60–68.

Mentesana, M., Rotz, G., Strang, D., & Swanick, M. (2017). 2017 Pharmaceuticals and life sciences trends. *Strategy& White Paper*. Retrieved from https://www.strategyand.pwc.com/trend/2017-life-sciences-trends

Muhtaroglu, F. C. P., Demir, S., Obali, M., & Girgin, C. (2013, October). Business model canvas perspective on big data applications. In Big Data, 2013 *IEEE International Conference on* (pp. 32–37). IEEE.

Otto, R., Santagostino, A., & Schrader, U. (2014). Raid growth in biopharma: Challenges and opportunities. *McKinsely Quarterly*. Retrieved from https://www.mckinsey.com/industries/pharmaceuticals-and-medical-products/our-insights/rapid-growth-in-biopharma

PWC. (2017). Pharma 2020: Supplying the future. Retrieved from https://www.pwc.com/gx/en/pharma-life-sciences/pharma-2020/assets/pharma-2020-supplying-the-future.pdf

SAS. (2010). Supply-chain analytics: Beyond ERP & SCM. *SAS White Paper*. Retrieved from https://www.sas.com/resources/asset/SAS_IW_FinalLoRes.pdf

Statista. (2017). *U.S. pharmaceutical industry – Statistics and facts*. Retrieved from https://www.statista.com/topics/1719/pharmaceutical-industry/

Waller, M. A., & Fawcett, S. E. (2013). Data science, predictive analytics, and big data: A revolution that will transform supply chain design and management. *Journal of Business Logistics, 34*(2), 77–84.

Wamba, S. F., Akter, S., Edwards, A., Chopin, G., & Gnanzou, D. (2015). How 'big data' can make big impact: Findings from a systematic review and a longitudinal case study. *International Journal of Production Economics, 165*, 234–246 Retrieved from https://doi.org/10.1016/j.ijpe.2014.12.031

Wang, G., Gunasekaran, A., Ngai, E. W., & Papadopoulos, T. (2016). Big data analytics in logistics and supply chain management: Certain investigations for research and applications. *International Journal of Production Economics, 176*, 98–110 Retrieved from https://doi.org/10.1016/j.ijpe.2016.03.014

Part II
Methodological Issues in Business Analytic

Chapter 6
Importance of Project Management in Business Analytics: Academia and Real World

Samir Shah, Alexander Gochtovtt, and Greg Baldini

Abstract Project management constitutes a powerful lever as organizations face increasing pressure to manage projects to budget, on time, and deliver more insights, in less time and with rapidly increasing amounts of data. This is critical especially in business analytics, with more than75% of organizations planning big data investments over the next several years. But the manipulation of massive amounts of data presents challenges – budgetary, time constraints, execution, proper manager skillsets, and such like. These challenges have cramped recent project rollouts, as only 37% of organizations have deployed big data projects in the past year; this suggests that filling the gap between data and insight remains a substantial hurdle as well as evolving need of project management for such projects. This chapter offers real-world examples of how project management professionals tackle big data challenges in a rapidly evolving, data-rich environment. Simultaneously, it establishes a bridge between business and academia as they both recognize the joint necessity to develop highly trained project managers to utilize the powerful and cutting edge analytical tools available to create value.

Keywords Analytics · Project management · Business analytics · Data science · Business intelligence · Agile methods

S. Shah (✉)
Drexel University, Philadelphia, USA
e-mail: Sns72@drexel.edu

A. Gochtovtt
Epsilon, Philadelphia, USA
e-mail: alexander.gochtovtt@epsilon.com

G. Baldini
Nihilent Technologies, Inc, Philadelphia, USA
e-mail: greg.baldini@nihilentinc.com

M. Anandarajan, T. D. Harrison (eds.), *Aligning Business Strategies and Analytics*, Advances in Analytics and Data Science 1,
https://doi.org/10.1007/978-3-319-93299-6_6

Introduction

Today, project management is a critical discipline in all areas of an organization, and these skills seem especially necessary in business analytics projects. Many studies and long experience have shown that managing business analytics projects can be particularly challenging; these projects are not much different in terms of project management challenges but often experience greater uncertainty and more change during project execution. According to 2013 Gartner research, more than half of all analytics projects fail to deliver the original features and benefits as well as running over schedule and budget.

Today's business analytics project managers need to develop a specific set of project management skills that help them become a stronger change agent, a coach, and an effective communicator to successfully manage and deliver projects (Viaene and Van den Bunder, 2011).

This chapter begins with the discussion of typical characteristics of business analytics projects followed by the importance of project management in business analytics projects. It also discusses the real-world examples from Epsilon – how project management tools and techniques are being used in Epsilon's business analytics projects. The last section of the chapter describes the efforts in academia (at Drexel University) to prepare business analytics students with project management skills.

Business Analytics Projects

As reported by Hendershot (2016), more than 75% of organizations have planned big data investments in the next 2 years, according to a 2015 Gartner report, yet only 37% of organizations have deployed big data projects in 2015, which suggests that filling the gap between data and insight remains a substantial hurdle as well as evolving need of project management for such projects.

This is part of a long and growing trend in business analytics. There is constant pressure to deliver more insight, in less time and with rapidly increasing amounts of data. The amount of analytical data has grown tremendously and continues to do so. Today, data comes from every device in an organization, rather than from a few privileged business applications (White, 2011).

The pressure to provide analysis based on these data is immense, with business decision-makers requiring ever tighter iterations to deliver insights; 64% of business managers report shrinking decision windows (White, 2011).

Business analytics projects are unique in many ways and typically face a variety of challenges primarily due to large amounts of data as well as uncertain requirements and outcomes. One of the primary differences between analytics projects and other IT projects is the focus on expanding understanding of data and the process it represents. This causes analytics projects to more frequently invite new inquiry as a result of successful execution. Other projects tend to be typified by more concrete knowledge and are a formalization of existing understanding or process. Analytics projects, when successful, cause individuals to understand a process in new ways – this learning prompts novel questions. This is a cyclical process.

Some analytics project challenges are:

- Analytics projects include much experimentation.
  - The value of answering a question is often unknown until the question is answered.
  - Late arriving data and requirements are the norm.
  - Often, discoveries in an early phase of a project materially alter the scope and timelines of later phases.
  - In short, questions beget answers, and answers beget new questions.
- Business analytics straddles the fence between business and technical stakeholders more than most IT development projects.
  - End users expect massive flexibility to leverage an analytics deliverable in ways not anticipated during its delivery.
  - Reporting logic changes often and is not the domain of the individuals performing technical development.
- There are myriad of alternatives to IT-driven solutions.
  - There is a proliferation of "phantom analytics" solutions – reporting and analytical artifacts developed by end users on their desktops in tools like Microsoft Excel, Microsoft Access, SPSS, R, and Python. These are often developed independently and without organizational support but are required to be managed or merged into a formal business analytics program.
  - Most line-of-business applications include reporting functionality that will offer different results to what might be developed externally.
- Business analytics projects have visibility across the entire organization.
  - It is not uncommon for the C-suite and low-level analysts to use the same analytical solution and reports and all levels of management in between.
  - Past failures can sour organizations to new projects.
  - Different functional areas may require access to the same solution, but have different expectations, leading to a "kitchen sink" mentality of including all possible functionality.
- Data challenges.
  - For analytical purposes, strong data consistency is often a requirement, but unfortunately, that is not the case across many IT systems.
  - Analytics projects must rely on available data. Insufficient data can reduce or eliminate the value of analysis.
  - Data volume and velocity present challenges in ingesting new and historical data

- Limited expertise.
  - Business analytics is a relatively young technical discipline; thus there are few individuals with expertise in the area relative to the massive demand for more insight. IBM projects the demand for data scientist to soar 28% by 2020 (364,000 job openings) in the United States (Columbus, 2017).

Project Management

Today, project management has become in demand within all areas of an organization – skills, such as scope and time management, cost and quality management, resource allocation and coordination, team building and communication, risk assessment, stakeholder and procurement management, project leadership, post-project assessment, and the ability to implement strategic change, among others, are sought by employers.

Project Management Institute (PMI – http://pmi.org), founded in 1969, is the leading association for the project management profession. Conducted since 2006, PMI's Pulse of the Profession report provides the major trends for project management now and in the future based upon their global survey of project management practitioners. According to their 2017 Pulse of the Profession 2017 report, due to poor project performance, organizations are wasting an average of $97 million for every $1 billion invested. At the same time, the project success rate is much higher (92% versus 33% of underperformers) for those organizations that are more mature with their project management practices.

There is an ongoing trend for business to increase the priority of analytics projects (Viaene and Van den Bunder, 2011). Additionally, analytics projects are often crosscutting within an organization and have highly variable scope and requirements. These factors combine to make project management a critical discipline to achieve success in analytics.

Typically, two methodologies are used to manage analytics projects: agile methods and CRISP-DM.

Agile Methods

Agile methods play a large role in business analytics projects. With the highly variable scope and requirements, it becomes necessary to have tighter integration of development and business stakeholders. Additionally, to justify the often large cost of business analytics projects, it is required to deliver value early. Finally, as the nature of interaction with an analytics solution is more free form and ad hoc, it benefits the business stakeholders to have early and frequent exposure to the data, the modeling process, as well as the solution deliverables. Overall, agile methods seem to provide benefits to business analytics projects (Kisielnicki and Misiak, 2016).

CRISP-DM

As mentioned in (McMahon, 2016) report, in analytics projects, five elements are critical at all phases of the CRISP-DM (cross-industry standard process for data mining) process. Project management ensures that the business stakeholders do not change the principal aim of the analytics outcome. As described in Project Management Institute's 2017 Project Management Body of Knowledge communication among the project stakeholders is one of the core areas of the project management methodology. Such communication should help a project in terms of clear understanding of the requirements, managing risks and changes as well as better outcomes as it goes through the five major phases of project management: initiating, planning, executing, controlling, and closing.

About Epsilon

Epsilon is a data-driven marketing company founded in 1969 and based in Irving, Texas (http://www.epsilon.com). The company provides database marketing solutions and services that integrate data, creative, technology, and customer experience for large brands across industries. Epsilon serves customers in automotive, financial services, healthcare, nonprofit, retail, telecom and technology, and travel and hospitality industries in North America, Asia Pacific, Europe, the Middle East, and Africa.

Some of its more analytical services include marketing services such as strategic and analytic consulting, marketplace assessment, competitive analysis, customer experience design, marketing mix allocation, investment justification, and predictive modeling and optimization.

Epsilon's technology teams execute small- to large-scale business and marketing analytics projects ranging from $50 K to $2 M in average engagement value using various technical (analytical) and project management tools and techniques. For the technical part, Epsilon uses COTS (common-off-the-shelf) tools such as Microsoft Project and PeopleSoft and statistical software suites such as SAS and R.

While some of the projects have a repeatable set of activities (i.e., CRM lift analysis, look-alike modeling, customer segmentations), different clients and a variety of data sources require bespoke project planning and cost accounting. The template activities required on a project are therefore managed by project managers using Microsoft Project, and high-level cost, budgets, labor, and burn rates are managed with enterprise tools including PeopleSoft. While Epsilon's project life cycle is PMI-based (not PMI-certified), the Advanced Consulting Group responsible for analytics project has significant experience in executing marketing analytics projects across various industries including finance and life sciences where analytical rigor is mandatory.

In large projects (typically project which are over 1000 total hours), Epsilon takes a multidisciplinary approach, as shown in Fig. 6.1, to partner strategy and analytics talent with its campaign execution team. The strategy team is responsible for asking and defining the key business questions related to execution, while the

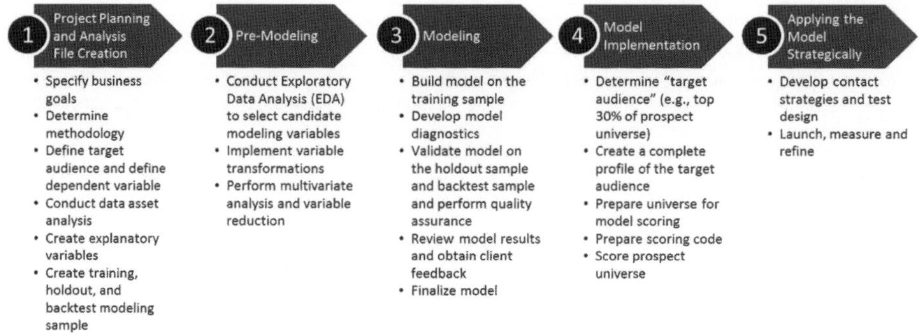

Fig. 6.1 Project management in analytics projects. (Source: Epsilon Data Management)

analytics team will define measurement standards and execute on the defined project that has been deemed to have enough practical utility for the business. Throughout this process, project management is essential for phasing the logical units of work properly across the execution team. Specifically, strategy is managed to provide quantifiable steps to the objective, while the analytics team iterates through the important steps of data collection, cleansing, analysis, modeling, and insights.

The model below (Fig. 6.1) has been useful in aligning clients, business representatives, and analytics subject matter experts to improve execution of projects. It does this by:

- Aligning business and analytical goals and objectives early
- Performing exploratory data analysis to understand data quality and assess project risks
- Obtaining client feedback during early model development and training
- Leveraging analytics SME to scale and apply the model to the business needs
- Testing in-market to provide client and business representatives with the proof points required to justify the effort and improve the business

This project approach has been especially useful with the increase in data-driven marketing which allows for a better understanding of audience needs, wants, and behavior using available client and third-party online data sources.

Typical Project Success and Challenges at Epsilon

Although many projects at Epsilon get completed successfully,they do face various challenges related to project management such as completing them on time and budget, lack of expertise in business analytics project management skills, and executing and interpreting results. Specifically, project management with business analytics projects falls into the following categories: (1) business risk, (2) data risk, (3) execution risk, and (4) staffing risks. Business risks are typically encountered when the business representatives do not have the full understanding of the analytical objectives or the ability of the project to help their business. Data risks are related to governance, quality, and data acquisition issues. Execution risk is often related to incomplete understanding of the stage gates and milestones required to ensure a project successfully moves forward, and staffing risks are usually related to skills, training, and availability of resources on the project. Successfully executed projects tend to have strong project managers who adhere to a methodology, who identify risk and mitigation strategies early, and who assess data quality and volume early enough to reduce budget waste.

Below is the description of two Epsilon projects where use of project management tools and techniques benefited the successful outcome.

Project 1: Life Science Project

Project's Business Goals and Objectives

A life science company was looking to reduce and optimize its marketing investment across multiple channels. They required a new way to go-to-market that reduced emphasis on traditional sales channels and increased the use of inside sales and digital sales support. The objectives were to define a weighted measurement framework to assess promotional response and optimize it based on the segments being targeted for promotion. The project spanned 5 months and involved a strong analytically focused core team along with varied business stakeholders. The core team consisted of executive level sponsors, several analytics subject matter experts (primary research, digital analytics, and statistical methods), and many business stakeholders who provided data and inputs on goals and objectives.

Analytics Tools Used

Standard analytical tools and techniques were used. From SPSS for primary research to SAS for latent class modeling, many work streams worked on improving data quality and business rules to drive a better integrated model.

Project Management Tools Used

Due to the highly matrixed nature of the organization and the executive sponsorship level, project management was very rigorous on the project. Since the project kicked off with several high-level stakeholder interviews, the business representatives were involved at the start and required frequent updates on the project status including complete key activities, risk assessment, and budget updates. Project meetings were scheduled twice a month, and tracking was done via Gantt charts and risk matrices.

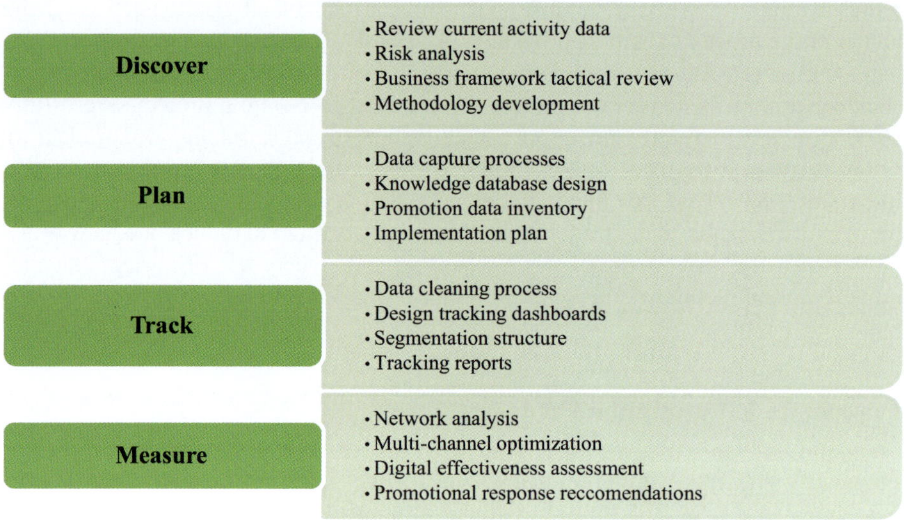

Fig. 6.2 Epsilon's integrated model

Key executives were interviewed to align on business objectives and goals, primary research was used where data was lacking, and data was aggregated, cleansed, and imputed where possible to create high-value data sets. Several work streams were undertaken simultaneously and combined to allow for an integrated model to be developed. The model, Fig. 6.2, was then trained and operated for several months before becoming the standard for the organization. The steps in this model are aligned to Stefan H. Thomke's four-step cycle: design, build, run, and analyze (Thomke, 2003).

The project helped commercial operations and marketing leaders better understand the digital preferences of their customers. This insight, in turn, allowed for an improved allocation of dollars to marketing tactics based on individual customer segments rather than channel allocations. It reduced some marketing waste and increased marketing operations yield.

Project Success and Challenges

The greatest success was establishing an enterprise-wide system for assessing and optimizing promotional response for the organization highest-value customers. The biggest challenge was obtaining organization buy-in to continue funding the project. The lack of an immediate deliverable in an organization with a medium analytics maturity meant constantly looking for low-hanging fruit to clearly demonstrate value and the practical usefulness of the work.

Lessons Learned

Rigorous project management was essential to align the different groups and to obtain buy-in for the project. Once all parties understood the potential benefit and quick wins the project would afford, momentum helped get the right resources applied to successfully complete the project. By having project management focused on delivering the solution and the analytics team focused on validating the solution for the business, the team was able to reduce organizational friction which could have caused the project to derail.

Project 2: Creative Performance Analysis Project

Project's Business Goals and Objectives

A specialty media advertising agency was looking to understand which aspect of their ad creative launched into market drove the highest level of performance. Experienced direct marketers have always executed tests to determine the optimal asset configuration to drive customer response; however, digital advertisers can be overwhelmed by the complexity of creative factors, media channels, and ad placements for similar analyses. The scope of the creative performance analysis project was to look for a way to understand what aspects of the creative drove digital performance and to what extent was the graphical choices material to performance.

Online/digital advertising, much like direct marketing, relies on various aspects of creative for effectiveness. While direct mail uses stamps, stickers, and other markings to invite customers to "act now, before the offer expires," digital ads use prominent fonts, calls to action (i.e., buttons in ads), and images to invite online users to click or submit information. Due to rich measurement ability in digital analytics, it is possible to determine the "effect" of various facets and levels related to an online ad using fractional factorial analysis performed either manually or using sophisticated testing tools such as Adobe Test and Target® or Optimizely®.

The project took a manual approach to extract the key attributes of several thousand online ads and attempt to link overall performance to these attributes. For example, would ads with people be more successful than ads without, or would ads with authoritative calls to action (i.e. "Click here") be more effective than those without (i.e., "More information?")? The business request came from senior business leaders, and the project was conducted over several weeks with a small team of subject matter experts. The project was managed by the chief analytics officer in close partnership with the analytics lead and subject matter experts. Additional support was provided by data analysts who extracted and compiled performance data related to the ads.

Analytics Tools Used

Microsoft Azure, Kafka, Tableau, Oracle, and Python were used to execute the project. Each of the aforementioned tools provided a unique and complementary function on the project: Microsoft Azure for its AI and cognitive services, Kafka for its dynamic data dashboards, Tableau for exploratory data analytics and reports, Oracle for performance data management and table calculations, and Python for log processing.

Project Management Tools Used

Project management was used to define and execute a proof of concept, to define a go-to-market approach with sales executive, to perform the analysis, and to rollout the project in order to support sales. The project was executed in a tight timeline and needed to be timeboxed to keep activities on track and scope managed.

Since the scope was flexible, the project manager created a well-defined project plan with deliverables, resources, and scope. Also, since the project was a prototype, scope was aggressively managed to focus on key deliverables including image assets collection, performance data tables, and dashboard design. Initial data collection and complexity around image assets caused several project delays which had to be addressed. Initially, the project called for Tableau to be used for dashboards, but due to compressed timelines, the analytics SME suggested Kafka to accelerate development time and keep the project on time.

Project Success and Challenges

Success was defining a minimum viable product which got instant market uptake after being shown to clients. While this project was helpful in defining a practical solution, the lack of stakeholder management during the project meant that scaling past the initial completion was difficult for the organization. No additional resources were applied, and the project was left as useful proof of concept.

How Academia Prepares Students with the Proper Skills

According to the Master's in Data Science's *25 Top Schools with Master's in Business Analytics Programs* report, a degree program in business analytics seems to be one of the growing majors within business schools. In addition to the foundation courses in business analytics, many such programs offer a set of electives, particularly, in project management, industry projects, and interdisciplinary programs.

For example, the Clarkson University's MS in Data Analytics program offers an elective course in Strategic Project Management; the University of Chicago's

MSc in Analytics offers a core course in leadership skills – teams, strategies, and communication – and the University of Minnesota's Master of Science in Business Analytics offers a core course, Project Management of Analytics Projects. Additionally, the University of Connecticut takes an interdisciplinary approach with its integrated MS in Business Analytics and Project Management, a program where students are trained with both advanced business analytics and project management skills.

Through such core, elective and interdisciplinary courses in project management, the universities aim to better prepare graduates with not only technical skills but also with the core project management skills to link between data analytics and business objectives which include managing projects within scope, time and budget, controlling and managing changes, understanding customer's needs, leadership, and managing project risks.

At Drexel University, both the undergraduate and graduate business analytics programs are offered by the Department of Decision Sciences and MIS where students are provided to learn and apply project management skills. This includes dedicated courses in the areas of project management and their practical applicability. Some of these courses also include practical classroom projects where students develop valuable project management-related insights from real-world perspectives.

Course Offerings

At present, we offer two courses dedicated to project management – MIS 361 (IS Project Management) (undergraduate level) and MIS T680 (Special Topic: Project Management) (graduate level). Both these courses introduce and explore the basic concepts and practices of project management that help students understand how the use of project management methodologies and tools can aid managers to plan and manage projects successfully.

These courses are structured around the key phases of a project life cycle – project initiation, project planning, project execution, project control, and project closeout. It also pays specific attention to the ten knowledge areas of project management as defined by the Project Management Institute's Project Management Body of Knowledge (PMBOK) – project scope, cost, time, integration, quality, communication, risk, human resources, procurement, and stakeholder management.

Additionally, the concepts of agile project management, international project management, and design thinking are also introduced. In the graduate version course, the related case studies and project management simulation – scope, resources, and schedule – from Harvard Business School are also discussed (Austin, 2013).

Business analytics students at Drexel can benefit learning from these project management skills and apply them to the business analytics projects. Table 6.1

Table 6.1 Project management skills and its applicability in analytics projects

Project management skills	Applicability in business analytics
Scope and time management	By definition, analytics projects are about learning new things and understanding business processes better. More than is typical in other technology projects; development leads to new questions and requests. It is critical to keep discipline around agreed-upon deliverables while balancing the need to answer current and relevant questions.Often the most critical questions to ask of data are unknown at the beginning of an analytics project
Stakeholder management	Analytics projects often involve a wide cross section of stakeholders, from line of business employees up to C-suite leadership. The wide range of interests and voices involved in defining the requirements of an analytics solution present challenges in managing expectations and balancing often conflicting needs
Strategic change management	Frequently, the goal of analytics projects is explicitly to identify new and better ways of doing business. By definition this changes the requirements of an analytics solution. Success in a project necessitates a future project to accommodate the results of the first project's success. This can be a virtuous or vicious cycle, and which it becomes depends on the ability of project leaders to react to changes they foment – change management is a core requirement of any successful analytics project
Team building and communication	Analytics projects often sit at a nexus of multiple source systems and groups of stakeholders. If a solution is to serve these groups effectively, it is necessary to be clear in communication among all stakeholders. Changes in a single source can ripple through a centralized analytical solution to affect many end users, who may or may not have any direct relationship to the source of a change. Similarly, instigators of changes may not realize how far the effects of their decisions might reach

describes how some important project management skills can be applicable in business analytics projects.

Real-World Project Offerings

In order to provide students real-world project management skills, an action learning multidisciplinary classroom projects are offered at Drexel as part of MIS 361 and another global classroom project course, MIS 347: Domestic and Global IS Outsourcing. These project problems are both domestically and globally based, ranging from working with profit, nonprofit, and government entities.

In some of these real-world problem-solving projects, students at Drexel typically collaborate with students from a major university in India as they develop the joint solution. The Drexel students are responsible for consulting with their customer – to understand the project requirements, to successfully communicate them to students in India, and to manage the overall project. The students in India are responsible for designing, coding, and testing the technical aspects of the business problem solution.

The purpose here is to help students in both countries develop valuable insights not only into how to produce joint solutions but to provide them opportunities to learn how to deal with project management-related challenges such as working with the team with radically different work patterns, negotiations and trust, cultural sensitivity, project requirement interpretations, time zone difference, risk management, quality control, and, more importantly, distant communication.

In many cases, students worked on these projects for 8–10 weeks to fully develop the solution using cutting-edge technology. In addition to deposit their individual work hours, students typically exchange many emails and text messages and conduct face-to-face video conference meetings and social media communication, a critical aspect of the project management.

Conclusion

It is critical to understand the experimental nature of discovery and development in an analytics project while balancing these loose requirements against organizational and political constraints to ensure that the experimentation drives toward a solution-oriented deliverable that can benefit the organization.

Project management provides mechanisms to overcome the typical barriers to the successful completion of these types of projects thereby unlocking the potential for knowing new and exciting things. Project managers with the right skills and experience can act as the backbone of complex analytics project, and this body of knowledge will become more and more important as data grows, analytical techniques become more complex, and business demand for such services increases.

The best systems, designed with perfect domain understanding, will still be worthless if the data flowing through them is insufficient or incorrect. Often analytics projects highlight or even uncover data quality and management issues, but it is important to understand and emphasize that this is not the same as creating those issues.

Project management professionals, more than ever, must be key facilitators and drivers of the delivery process. Stefan Ahrens from SAS recommends that so long project manager pay attention to the critical success factors of analytics projects such as allow time to build domain expertise, limit the scope and try to be realistic, be aware of the iterative nature of analytics project, and allow time to fix data quality and data management issues, there should not be any reason why analytics projects should not benefit from project management (Ahrens, 2014).

References

Ahrens, S. (2014). What an IT project manager should know about analytics projects. SAS Voices. Retrieved from https://blogs.sas.com/content/sascom/2014/09/15/it-project-manager-and-analytics-projects/

Austin, R. D. (2013). Project management simulation: Scope, resources & time. Harvard Business Publishing Education. https://cb.hbsp.harvard.edu/cbmp/product/4700-HTM-ENG

Columbus, L. (2017). IBM predicts demand for data scientist will soar 28% by 2020. *IBM White Paper*. Retrieved from https://www.forbes.com/sites/louiscolumbus/2017/05/13/ibm-predicts-demand-for-data-scientists-will-soar-28-by-2020/#6519b2b97e3b

Gartner. (2013). Gartner predicts business intelligence and analytics will remain top focus for CIOs through 2017. *Gartner News Article*. Retrieved from http://www.gartner.com/newsroom/id/2637615

Hendershot, S. (2016). Data done right: Learning to target the right type of data can uncover insights that drive project success. *PM Network, 30*(3), 40–45.

Kisielnicki, J., & Misiak, A. M. (2016). Effectiveness of agile implementation methods in business intelligence projects from an end-user perspective. *Informing Science: The International Journal of an Emerging Transdiscipline, 19*, 161–172.

McMahon, A. (2016). All assumptions are false! 7 lessons I wish I paid more attention to on every predictive analytics project. Presidion White Paper. Retrieved from http://www.presidion.com/wp-content/uploads/2016/03/1603_AM_7Lessons_PA_Project.pdf

Project Management Institute. (2017). Project Management Body of Knowledge (PMBOK). 6th Edition. PMI Website. Retrieved from https://www.pmi.org/pmbok-guide-standards/foundational/pmbok/sixth-edition

Thomke, S. (2003). *Experimentation matters: Unlocking the potential of new technologies for innovation* (pp. 97–98). Boston: Harvard Business School Press.

Viaene, S., & Van den Bunder, A. (2011). The secrets to managing business analytics projects. *MIT Sloan Management Review, 53*(1), 65–69.

White, D. (2011). Agile BI – complementing traditional BI to address the shrinking decision-window. Aberdeen Group White Paper. Retrieved from https://www.montage.co.nz/assets/Brochures/Aberdeen-Agile-BI.pdf

Chapter 7
A Review and Future Direction of Business Analytics Project Delivery

Deanne Larson

Abstract Business analytics is a core competency critical to organizations to stay competitive; however, many organizations are challenged at business analytics delivery, and these projects have a high rate of failure. The volume, variety, and velocity of the big data phenomenon and the lack of current methodologies for delivering business analytics projects are the primary challenges. Applying traditional information technology project methodologies is problematic and has been identified as the largest contributing factor for business analytics project failure. Business analytics projects focus on delivering data insights as well as software delivery. Agile methodologies focus on the ability to respond to change through incremental, iterative processes. Agile methodologies in software delivery have been on the rise, and the dynamic principles align with the discovery nature of business analytics projects. This article explores the big data phenomenon, its impact on business analytics project delivery, and recommends an agile framework for business analytic project delivery using agile methodology principles and practices.

Keywords Agile methodologies · Analytics projects · Big data · CRISP-DM · Agile software development

Introduction

Deriving business knowledge and value from data has never been more critical to organizations to stay competitive nor has it been more challenging due to the disruptive nature of big data (Abbasi, Sarker, & Chiang, 2016; Bole, Popovič, Žabkar, Papa, & Jaklič, 2015; Davenport, 2013, 2015; Sim, 2014). The big data phenomenon – the volume, variety, and velocity of data – has created new challenges for organizations to deliver business analytics (Abbasi et al., 2016; Davenport, 2013). One primary challenge is using existing project methodologies to deliver business

D. Larson (✉)
Larson & Associates, LLC, Seattle, WA, USA

M. Anandarajan, T. D. Harrison (eds.), *Aligning Business Strategies and Analytics*, Advances in Analytics and Data Science 1,
https://doi.org/10.1007/978-3-319-93299-6_7

analytic projects. Applying traditional project methodologies is problematic and has been identified as the largest contributing factor for business analytics project failure; organizations are treating business analytics projects like any other information technology (IT) project (Demirkan & Dal, 2014). New agile processes are needed to increase business analytics project success (Davenport, 2013; Demirkan & Dal, 2014; Gartner Research, 2015).

Agile methodologies focus on the ability to respond to change through incremental, iterative processes and were derived from software development (Aston, 2017). The manifesto and principles for agile software development (ASD) were published in 2001, and since then, the objectives and principles have been interpreted and applied to IT project delivery. Beck et al. (2001) outlined the core values of the manifesto: individuals and interactions over processes and tools, working software over comprehensive documentation, customer collaboration over contract negotiation, and responding to change over following a plan. The result of following these values is that project delivery becomes less formal, more dynamic, and customer-focused. Although some organizations may not classify business analytics projects as IT projects, these initiatives rely heavily on the need for information and technology to be successful (Davenport, 2013). This paper explores the application of agile methodologies and principles to business analytics project delivery.

The Changing Data Landscape

The increased use of internet-connected smart devices has altered how organizations and individuals use information (Abbasi et al., 2016; Davenport, 2013; Halper, 2015). The Internet of Things (IoT), where data collection is embedded into devices, results in exponentially increasing amounts of data and contributes to the demand for fresher data (Halper, 2015). Exploring the change in the volume, variety, velocity, and veracity of data, otherwise known as the characteristics of big data, demonstrates the challenge to traditional delivery methodologies. Business analytics, in several ways, is a result of big data, primarily due to the need to analyze data (other than traditional structured data) such as text, streaming, machine-generated, and geospatial data (Halper, 2017). Big data and business analytics go hand in hand; thus project methodologies used in business analytics need to consider both (Davenport, 2013; Demirkan & Dal, 2014; Gartner Research, 2015).

Volume

The volume of data created used to be described in terabytes; however, terabytes have now been replaced with petabytes and exabytes. The growth of data impacts the scope of data used in projects. Scope increases project complexity where new technology is used to accommodate more data and more data results in increased data quality issues (Abbasi et al., 2016).

Variety

Data variety becomes a concern for project delivery as the types of data sources to be used for analysis increase. The variety of data means increasingly complex forms of data such as structured and unstructured data (Abbasi et al., 2016; Davenport, 2013; Halper, 2015). Traditionally, structured data is created in rows and columns and easily understood; however, unstructured data comes in different forms, levels of details, and without clear metadata, complicating the ability to understand and use (Abbasi et al., 2016; Davenport, 2013; Halper, 2015).

Transactional data, images, social network content, sensor data from IoT, rich text, and clickstream are all examples of the variety of data used in analysis. Additionally, data sources are not only valuable independently but also integrated, resulting in the need to integrate a variety of data structures. Both the translation and integration of data sources increases the complexity of projects.

Velocity

Velocity focuses on the speed of data creation. Using Twitter as an example, in 2014, one billion tweets were created on average every 3 days (Abbasi & Adjeroh, 2014). Data velocity challenges the ability to analyze trends and patterns. Historically, fresh data may have been categorized as data from the previous business day; but in the case of IoT, data that is an hour old may be too old for analysis (Halper, 2015). Data acquisition becomes a challenge as traditional data acquisition focused on extract, transformational, and load (ETL) of data. Increased velocity changes the order where data is loaded first, then analyzed, otherwise known as extract, load, and then transformation (ELT) (Abbasi et al., 2016, Davenport, 2013; Halper, 2015).

Veracity

Veracity deals with data accuracy. Projects that derive value and knowledge from data have always been challenged with credibility and reliability issues. Volume, variety, and velocity complicate this more, and veracity problems increase. Understandability of unstructured data is a challenge as metadata does not exist and data completeness varies. The different types of sources complicate the veracity of data, and the complexity of deriving knowledge from data increases (Abbasi et al., 2016; Davenport, 2013; Halper, 2015).

Traditional methodologies focus on understanding requirements first, using requirements to derive the design, design to complete development and testing, resulting in the deployment of software. Big data requires ingestion of data first to explore data, understand the data, then deriving requirements, changing the traditional sequence of activities for project delivery. Projects that include big data cannot be treated as a typical IT project, and project failures have provided insight on how practices should change (Demirkan & Dal, 2014).

Best Practices in Business Analytics Project Delivery

Business analytics is still evolving; thus best practices are just starting to emerge. The following best practices have emerged from leading analysts and practitioners based on some of the failure points of business analytics projects (Demirkan & Dal, 2014; Gartner, 2015; Halper, 2017; Larson & Chang, 2016). These best practices reinforce the failure of existing methodologies to address the challenge of big data and the need for leveraging agile practices.

Due Diligence in Defining Business Goals and Problem Statement

Business analytics projects have failed due to the lack of clear business goals or problem statements. The success and value of a business analytics project will be defined at the start of the project (Demirkan & Dal, 2014; Gartner, 2015; Halper, 2017; Larson & Chang, 2016). By clearly defining the business goals to be addressed and formulating a problem statement, the scope of the project becomes clear. A clear scope impacts the steps of data understanding and data preparation. If the investment is not made up-front to understand expectations, subsequent work on the business analytics project is impacted. The problem statement should clearly identify the data needs of the project (Gartner, 2015; Halper, 2017; Larson & Chang, 2016). Leveraging a collaborative team of stakeholders and delivery personnel can assist in addressing this.

Allow Time for Data Understanding, Acquisition, and Preparation

A key failure point in business analytics projects is allowing for the time needed to get the data required for the project. Often the time needed to work with the data is underestimated or not known at the start of the project. The challenges that come with big data emerge in this area. Working with multiple technology platforms, multiple data structures, data sampling, data integration, and merging data into a final set for modeling takes time and planning. Additionally, once a final data set is created, the project team should document the data lineage to ensure the data set logic is clear and the data set can be recreated via a data pipeline (Gartner, 2015; Halper, 2017; Larson & Chang, 2016).

Identify Needed Toolsets and Skill Sets at Project Start

As outlined, the technology infrastructure in the business analytics industry can be varied and complex. Tools can be open-source where support is lacking adding to the complexity. Technical tools are needed for analytical modeling and data wrangling which are key for the success of the project. The project team will need to have a diverse set of roles focusing on technology, analytical modeling, statistics, and business skills. The project leader will need to have a strong technical and analytical skill set to lead the project. Identifying the tools and skill sets at project start prevents technical and team barriers during the iterative stages (Gartner, 2015; Halper, 2017; Larson & Chang, 2016). Leveraging small versatile teams can be of benefit here.

Allow Time for the Cycle of Modeling and Evaluation

The value of business analytics occurs when insight is attained. Insight refers to the capabilities provided by the model which typically falls in the categories of lowering risk, increasing revenue, increasing productivity, and supporting and shaping an organization's strategy (Larson & Chang, 2016). Modeling and evaluation tends to be experimental which results in answers but also more questions. Models also need to be built with different algorithms to be able to validate accuracy increasing the time needed for more experimentation (Gartner, 2015; Halper, 2017; Larson & Chang, 2016). Iterative software development cycles can be leveraged to support the cycle of modeling and evaluation.

These best practices highlight the need to consider adapting agile principles in business analytics delivery. The next section explores the prevailing methodology and agile application in business analytics.

Methodology

Methodologies have value in that they help deliver projects effectively and efficiently, which, in turn, enables business value from the project investment (Grimes, 2006; Sim, 2014). Business analytics projects often are initiated without clear objectives and outcomes, inviting constant scrutiny on whether business value occurs (Larson & Chang, 2016). Deriving value from information means that each business analytics project needs to deliver usable results, where each project is a transformational effort to create knowledge from data (Grimes, 2006). The scope of each business analytics project covers all activities that need to occur in the information value chain, which includes converting data into information and information into knowledge (Abbasi et al., 2016). Business analytics projects have a high degree of complexity as converting information into knowledge requires the use of statistical and analytical models as well as incorporating the use of big data (Halper, 2015). The Cross-Industry Standard Practice for Data Mining (CRISP-DM) is the top methodology in use in business analytics (Piatetsky, 2014).

CRISP-DM

CRISP-DM is a data mining process model that conceptually describes the stages that are used to tackle data mining problems. CRISP-DM was originally created to align with data mining, but it has organically evolved into the primary approach used by data scientists (Piatetsky, 2014). CRISP-DM contains six stages which appear to be in sequence; however, the stages are not strictly sequential, and iterating through the stages is expected (Marbán, Mariscal, & Segovia, 2009).

Business Understanding

Business understanding focuses on determining the business requirements and objectives to create a problem definition. Business analytics initiatives start with a question of interest or problem to be addressed. The outcome of this stage is a problem definition. The problem definition may be diagrammed using decision modeling or other approach (Marbán et al., 2009).

Data Understanding

Once a problem statement is understood, data collection can proceed. Data collection involves obtaining data attributes required to address the problem statement. Often data integration is required as data may come from various sources in various formats. Once data is useable, data is profiled and statistically analyzed to determine demographics, relationships, distribution, and quality levels. The outcome of this stage is initial insights into the data to support the problem statement (Marbán et al., 2009).

Data Preparation

Data preparation is necessary to create the data set that will be used in the analytical modeling stage. Data preparation includes the final integration of various attributes, cleansing of attributes, and deriving of new attributes. Activities in this stage are often iterative as new data may be required to support the subsequent modeling and evaluation stages. The outcome of data preparation is a data set that is to be used in the first iteration of modeling (Marbán et al., 2009).

Modeling

Various modeling techniques and algorithms are explored in this stage as there are several techniques that can be used to address one problem. The parameters of the analytical models are adjusted to improve model performance. It is possible to go back to the data preparation phase if additional data is needed or formatting needs to be adjusted (Marbán et al., 2009).

Evaluation

Models are evaluated based on minimizing accuracy errors, bias, variance, and overall fit to the business objectives or problem statement. Before moving to deployment, a final review is completed to confirm the steps taken to create the model, then a final decision is made move to the deployment stage (Marbán et al., 2009).

Deployment

The model deployment scope will depend on the frequency of the model. Models can run one time creating descriptive analytics or be a repeatable scoring model. Deployment to a production environment often occurs at this stage. Models are monitored to ensure quality as model accuracy can degrade due to change and time (Marbán et al., 2009).

CRISP-DM was created for data mining in 1996 and does not highlight the granularity of effort needed for successful business analytics outcomes (Piatetsky, 2014). According to Piatetsky (2014), CRISP-DM as a framework is valuable, but it lacks the detail needed for business analytics teams to address the challenges posed by big data and modern business analytics. There was an effort to create CRISP-DM 2.0 in 2007, but there is no new research or activity in this area (Piatetsky, 2014). The original web site for CRISP-DM is no longer active, and IBM is the only organization still including CRISP-DM in its system documentation (SPSS Modeler). CRISP-DM does not directly contribute to business analytics project failure; however, it does not provide a modern methodology that can be leveraged for business analytics project success (Piatetsky, 2014).

Agile Methodology

Keith (2006) recognized that agile methodology, although based on software development, can be applied more broadly in projects. Many agile methodologies have emerged since the agile manifesto was created in 2001. Methodologies such as

Scrum, Kanban, Lean Development, and Extreme Programming (XP) are examples, and each shares the same properties of focusing on people and results through the use of collaboration and streamlined methods (Alnoukari, 2015). While there are benefits in each of these methodologies, the values and principles of agile development versus the specific methodology are really the focus (Keith, 2006). Beck et al. (2001) created the values and the principles that form the basis for agile methodologies.

Application of agile methodology principles means maximizing information flow between team members, and reducing the time between decisions, resulting in speed and flexibility. Agile is not rigid but adaptable (Keith, 2006). Methodologies such as Scrum, Lean Development, and XP are adaptive approaches but use a plan-build-revise lifecycle. Adaptive Software Development is another agile method that embraces an adaptive approach, where plan-build-revise is replaced with speculate-collaborate-learn (Alnoukari, 2015). Adaptive Software Development demonstrates the applications of agile principles versus a prescriptive methodology. Keith (2006) outlines that plan is too rigid, build is too process-focused, and revise does not address the learning that occurs in the process and speculate-collaborate-learn is more agile.

The agile movement has also produced agile project management which focuses on scope flexibility, collaboration, and delivering quality results (Layton & Ostermiller, 2017). Agile project management was also built on the agile principles and used Scrum as a framework, XP for a focus on quality, and Lean Development, to reduce rework. Agile project management focuses on results, communication, people, and flexibility.

Ultimately, the goal of business analytics is to produce a data product, which is a technical asset that ingests data to return algorithmically generated results. An example would be a recommendation engine that tracks and ingests user's preferences and makes recommendations. Data products are designed to integrate into core business applications and processes to enable smarter business decisions at all levels.

The concept of the data product provides a basis of why agile methodology principles should be combined with CRISP-DM processes. Keith (2006) outlined that working software is the measure of agile project success, but in business analytics, the working software is not valuable unless it provides business insight and better business decisions. The next section proposes a practical framework for business analytics project delivery leveraging both CRISP-DM phases and agile principles, tools, and roles.

Practical Agile Framework for Business Analytics Project Delivery

Sahu (2016) proposes existing data mining methodologies need to be more dynamic in order to increase business value enabled from business analytics projects, and by adopting agile principles, this could be possible. ASD has been applied to IT projects to deliver solutions faster with an intent on increasing business value (Aston, 2017). Alnoukari (2015) presented a methodology that leverages ASD principles and the CRISP-DM process called ASD-BI (agile software development-business intelligence) supporting that both CRISP-DM and ASD can be combined to improve project delivery and business value.

The practical agile framework focuses on using the CRISP-DM phases to guide the business analytics project. Business understanding focuses on determining the business requirements and objectives to create a problem definition. Data understanding involves obtaining data attributes required to address the problem statement. Data preparation is necessary to create the data set that will be used in the analytical modeling stage. Modeling focuses on exploring modeling techniques and model creation. Evaluation is where models are evaluated based on minimizing accuracy errors, bias, variance, and overall fit to the business objectives or problem statement. Last, deployment focuses on moving a model to a production environment for business use.

Within each of the phases, best practices are proposed based on emerging research, agile values are aligned with the phases highlighting which principles and values to focus on, tools are recommended that align with emerging best practices, agile roles are called out to leverage, and outputs are defined that assist project leaders in planning deliverables. Phases of data understanding, data preparation, and modeling are called out to be iterative. The full framework is depicted in Fig. 7.1. The value of the practical agile framework is fourfold: it provides best practices that

Phase	Business Understanding	Data Understanding	Data Preparation	Modeling	Evaluation	Deployment
Activities	Determine Business Objectives Assess Situation Determine Project Goals Produce Project Plan	Collect Initial Data Describe Data Explore Data Verify Data Quality	Select Data Clean Data Construct Data Integrate Data Format Data	Select Modeling Techniques Generate Test Design Build Model Assess Model	Evaluate Results Review Process Determine Next Steps	Plan Deployment Plan Monitoring and Maintenance Produce Final Report Review Project
Best Practices	Problem Modeling Technology Inventory	Data Profiling Quality Reporting	Feature Selection Repeatable Pipeline and ETL	Model Train Model Test	Model Comparison Performance Testing and Tuning	Production Monitoring for Effectiveness
Agile Value	Collaboration; Insights	Collaboration; Discovery; Iteration	Collaboration; Discovery; Iteration; Working Software	Collaboration; Discovery; Iteration; Working Software	Collaboration; Discovery; Iteration; Working Software	Working Software
Tools	User Stories Cause and Effect Diagrams Project Charter Release Planning	Product Backlog Burndown Chart Daily Meetings Data Profiling Modeling Tool	Burndown Chart Daily Meetings Pipeline Creation ETL Modeling Tool	Burndown Chart Daily Meetings Modeling Tool	Burndown Chart Daily Meetings Modeling Tool	Retrospective Meeting Monitoring Dashboard
Team Roles	Project Lead Product Owner	Project Lead Product Owner Project Team Data Scientist	Project Lead Project Team Data Scientist	Project Lead Project Team Data Scientist	Project Lead Product Owner Project Team Data Scientist	Project Lead Product Owner Project Team Data Scientist
Outputs	Project Charter	Data Quality Report Candidate Features	Working Software (ETL) Train and Test Data	Working Models Test Results	Final Model (Data Product)	Deployed Model Monitoring Processes

Fig. 7.1 Practical agile framework

prepare for the challenges of big data; it provides an alignment with agile tools and phases where practitioners can leverage these from known agile methodologies; it highlights the agile roles that contribute to the outputs of each phase; and provides more detail to support the application of CRISP-DM by practitioners with identified practices, tools, roles, and outputs. The combination of these four objectives also balances the need for working software and delivering insights which are the primary focus of business analytics projects.

Framework Explanation and Value

Activities

The framework activities align with the activities completed with each phase of CRISP-DM. These activities are leveraged due to prominent use of CRISP-DM as a driving methodology.

Best Practices

The best practices suggested are derived from documented failure points and emerging best practices. These best practices include:

- Problem modeling such as using a cause and effect diagram will help to visually display the many potential causes for a specific problem or effect, and it is very useful in investigating and clarifying the problem or opportunity for the business analytics project.
- Technology inventory includes determining what existing technology can be leveraged and what technology is needed. Addressing early in the life cycle ensures no training curves or technology integration issues in later phases.
- Data profiling is the process of examining the data from an existing data source and collecting statistics or demographics about that data. Data profiling tools expedite the data understanding stage and provides data quality reports. Data profiling also helps establish the metadata for the project.
- Quality reporting is the data quality reports derived from data profiling.
- Repeatable pipeline and ETL is a service that is used to process and move data between different storage platforms. Data pipelines enable easier data ingestion and are reusable. Extract, transform, and load tools enable easier data cleansing, reformatting, merging, and integration for training and test sets. ETL tools can also support metadata and data lineage collection.
- Model train and test data include the data sets to train and test the model. Ensuring that these data sets are produced appropriately using the correct sampling technique, and cross validation expedites the modeling phase.

- Model comparison refers to having additional models using different techniques for comparison to further validate model accuracy.
- Performance testing and tuning focuses on the performance of the data product and ensure working production-ready software in the deployment phase.
- Production monitoring for effectiveness includes creating monitoring dashboards to track the performance and effectiveness of analytical models in production. Monitoring dashboards assist in tracking the business value of the analytical model.

Agile Values

To illustrate the ASD values in the framework, a short description for each value was created. Aligning the values to each stage highlights where the ASD values contribute to the outputs from each stage.

Tools

Agile methodologies have introduced valuable tools to improve communication and increase business value. Some examples include user stories, daily meetings, and retrospective meetings.

Team Roles

Team roles have been identified based on common roles used in agile methodologies and software development projects. Roles adopted from agile methodology include project lead and product owner. The project lead enables the project team to be productive and tracks the progress and success of the project. A project lead facilitates and removes barriers to delivery, as an example. The product owner is an agile methodology role that represents the primary stakeholders and is responsible for maximizing the value of the product produced – in this case the data product. A product owner is important to ensure the data product produced addresses the business problem at hand.

Output

The outputs in the framework are the tangible artifacts created in a stage. Focusing on producing a minimal number of key artifacts that provide value allows the project team to be the most efficient in each stage.

The next section focuses on the application of the Business Analytics Project Delivery Framework.

Framework Example

The framework outlines practices and principles which improve business analytics project delivery. It has been used by the author as part of a consulting practice in over 20 different business analytics projects in different industries. Time to value has increased (reduced time) 20–40% since the first use. Improvements in stakeholder expectations, data acquisition time, reducing technology complexity, deployment ease, and measuring business value after deployment are the noted areas of success.

The framework was used on a case of optimizing ordering and delivery of a product line for small convenience stores. The outcomes and elapsed times tracked for each stage demonstrate an end-to-end delivery time of 30 days or a 28% improvement over a similar project (similar scope and effort) at the same organization. The data products produced were monitored for 90 days, and the result was a 25% improvement in store service levels. The application of the framework to this case follows.

Business Understanding Phase

The case studied focused on how to optimize ordering and service delivery schedules for a product line sold in small convenience stores in the USA. The primary scope of the project was to provide service schedule optimization for small stores to increase store service excellence. Product line ordering and restocking responsibilities belong to the manufacturer and not the convenience store; thus the convenience stores are the customer in this case.

This problem statement was further broken down into two problem/opportunity statements: improve prediction for store-level days of supply and recommend store-level service frequency for upcoming periods to be used by district managers for service frequency planning.

Best Practices Used

The focus of the business understanding phase is to have a complete view of the problem. The primary problem reported was that convenience store complaints had risen, and popular products were often either overstocked or understocked. Additionally, deliveries were inconsistent or late. Problem modeling sessions with convenience store owners, delivery personnel, and district managers produced the two problem/opportunity statements. These statements enabled a clear scope for the project, provided insight into the data that would be needed, and identified the potential analytical modeling techniques to be used. The technology inventory was also reviewed for potential gaps.

Tools Used

Cause and effect diagrams were used to investigate the convenience store complaints and model root causes. Root cause identification was the key to formulating problem statements. User stories were also leveraged to better understand the customer experience and the desired outcomes.

Time to Complete

The time to complete the business understanding phase was 8 business days where all of the activities listed in this phase of the framework were completed. The deliverables of this phase included the problem statements, project charter, and identified technology gaps. A similar case at the same organization took approximately 11 business days to complete business understanding activities and establish problem statements.

Data Understanding

The problem statements revealed the data sources to be included in scope which included customer data (convenience store profiles), visit data (delivery schedule information), delivery data (product levels, order levels), and sales data. The primary focus of the data understanding stage was to acquire the data sources, complete exploratory data analysis which includes data quality evaluation, and identification of candidate features. The outcome of this phase was the data quality report and the candidate features list.

Best Practices

The primary best practice leveraged in the data understanding phase was data profiling. Data profiling is the process of examining the data from an existing data source and collecting statistics or demographics about that data. Data profiling tools expedited exploratory data analysis and reduced the amount of time by 50% over previous initiatives. Data quality reports produced by the data profiling tool provided data demographics which enabled quality issues to be quickly identified. Additionally, the reports provided summary statistics and a correlation matrix to quickly identify and evaluate candidate features.

Tools Used

Data profiling provided the most benefit in this phase as it reduced discovery time. Other tools leveraged included short daily meetings, sometimes referred to as stand-up meetings in agile, which reduced the back-and-forth communications between data scientists, project leads, and project owners. Data acquisition, storage, and integration were tracked via a burndown chart. The development occurring in this phase was used as the basis for the ETL and data pipeline.

Time to Complete

The time to complete this phase was 8 business days, which was a 50% improvement over a similar project. The application of data profiling provided the largest gains from the framework by expediting the exploratory data analysis and data quality evaluation and reducing the number of iterations required. The candidate features identified in this phase were customer demographics, delivery frequency, store sales, and elapsed time between deliveries.

Data Preparation

The time spent on data preparation is highly dependent on a successful data understanding phase. The data preparation phase focused on producing the sample data to be used in modeling. Data cleansing and transformation were completed to produce a sample size of 500,000 records, representing store features over the last 6 months. The development work completed in the data understanding phase was leveraged for data preparation.

Best Practices

The best practices applied in data preparation focused on producing the required sample data and developing a data pipeline to be leveraged in production. Final feature selection was completed based on the analysis provided from exploratory data analysis and the data quality report. Producing a large quality sample provided a base of data to use for multiple modeling iterations and prevented rework in the modeling phase. A focus on producing a working data pipeline saved development time in the evaluation and deployment phases.

Tools

Short daily meetings continued to be leveraged to keep communication flowing between data scientists, project leads, and project owners. The development turned more formal in this phase focusing on producing working software for the ETL and data pipeline. Burndown charts were used to track development work. The early versions of working software were used to create the sample file of 500,000 records.

Time to Complete

Most of the time gains in data preparation were contributed to the thoroughness of the exploratory data analysis and data quality work completed in the data understanding phase. Less rework was required to finalize the features list and build the cleansing and transformation to produce the final data set. The time to complete this phase was 5 business days compared to 7 business days of a similar project.

Modeling

Based on the problem statements, the prediction targets were identified in the business understanding phase. The prediction targets were the optimal days of product supply and optimal number of deliveries per store. The modeling approach chosen was linear regression, and two different nonparametric modeling techniques were used for regression to expedite the evaluation phase. The modeling phase also leveraged the 500,000 record sample file produced in the data preparation phase. The outcome of this phase was four models – two different approaches for two problem statements.

Best Practices

The improvement observed in the modeling phase was directly dependent on a high-quality sample file which was split into smaller train and test files. Cross validation was used to complete multiple training and testing iterations. Using two different modeling techniques enabled a baseline comparison for modeling performance.

Tools

Daily meetings continued to be leverage, and worked continued on the ETL and data pipeline to prepare the software for production. Modeling tools were used to produce two complimentary models for each problem statement. Modeling tools needed to have a robust set of analytical functionalities, such as having multiple approaches to linear regression, to ensure no time is lost in this phase.

Time to Complete

A trend can be observed where improvement gains in subsequent phases are a result of improvement of prior phases. Most of the gains in the modeling phase are attributed to the high-quality sample file and clear problem statements. Time lost in the modeling phase in a similar project was due to the need to recreate the sample file and create new features. The time to complete this phase was 5 days compared to 8 days in a similar project.

Evaluation

Model evaluation focused on the performance comparison of the two regression models for each problem statement. Both models focused on predicting the optimal supply days and delivery frequency for individual stores. Stores were subjected to standard supply days and delivery schedules which were not geared to store characteristics. The outcome of this phase was the final recommendation of models.

Best Practices

The efficiencies gain in this phase were dependent on having comparative models. Each model had its own accuracy levels; however, accuracy levels and best performing models cannot be chosen without another to compare too. Using two different regression approaches (decision tree and ensemble methods), time to decision approval was shortened. The data pipeline worked continued to ensure a production-ready working software for the deployment phase.

Tools

Daily meetings and the use of burndown charts continued to support the data pipeline development. A final meeting to present findings to business stakeholders was conducted to gain approval for the final models.

Time to Complete

The evaluation phase took 2 business days. The improvement gains were directly attributed to having a minimum of two models to compare for each problem statement.

Deployment

Two models were recommended in the evaluation phase. At this time in the project due to early development efforts, working software for the model and the data pipeline were ready for production. This included the software needed to create the data structures, keep the data fresh, and run the models weekly. The outcome of this phase was deployed software and a dashboard to monitor model performance.

Best Practices

Producing production-ready software and creating a dashboard that was leveraged for production monitoring of the models were the best practices. Working on the software development in parallel to the modeling and evaluation phases helped prepare for the deployment phase.

Tools

A retrospective meeting to review the lessons learned and complete a final walk-through of software readiness was conducted. Lessons learned were documented by the project lead, and the product owner worked with the IT team to promote the final code to the production environment. The IT team received a package with the model, pipeline, and dashboard code to deploy.

Time to Complete

The end-to-end time to complete deployment was 2 days which included the final meeting and software package hand off to IT. The deployment time in a similar project was 7 days due to the lack of production-ready software at this phase. In the comparison project, developing production-ready software was not addressed until a final model was approved, delaying time to value.

The total time to complete this project was 30 days with an improvement of 8 days over a similar project or a 28% improvement in delivery time with the application of the framework. Some of the lessons learned from the application of the framework outlined the challenges of applying ASD directly to business analytics. These challenges encountered helped evolve the framework to its current state and are explored in the next section.

Challenges of ASD Applied to Business Analytics

Applying ASD principles to business analytics projects identified some challenges including time-boxing, the discovery nature of data wrangling, and the focus of software delivery versus insights derived from business analytics.

Time-Boxing

Time-boxing means that time boundaries are placed around an activity and are common practice in agile methodologies. The time-box becomes a constraint that drives the team to focus on value or minimal viable product (MVP). One of the challenges that comes with using a time-boxed approach is that business analytics is a process that uses data to follow a chain of analysis. In this chain of analysis, it may not be clear what the next question is until the current question is answered. In addition, the next area of analysis may be clear, but there may be a dependency on gathering or acquiring more data. Given the primary outcome of actionable insights from business analytics, time-boxing does not promote iterative and progressive analysis.

Data Wrangling

Data wrangling is an umbrella term used to describe the process of acquiring, transforming, cleansing, and integrating data into one "raw' form that enables the consumption of data (Jurney, 2017). The CRISP-DM stages of data understanding and data preparation require multiple iterations of data wrangling before the modeling stage can begin. Data wrangling also pushes up against constraints of the time-box. ASD focuses on the client, the technical team, and the business, but less value is placed on the data. In business analytics, all of the value is derived from the data.

Insight Versus Activity

In Scrum, a popular agile methodology, work items will be assigned to a developer during a sprint, which results in software. Again, the focus with ASD is working software; however, working software may produce no insight. Without the insight, business analytics produces no value. Multiple iterations are needed to produce insight. Jurney (2017) refers to the multiple iterations in business analytics as concurrent experiments. ASD focuses on the completion of activity, where business analytics is focusing on actionable insights.

Quality, Testing, and MVP

In ASD, testing is a component of each development phase accomplished via constant feedback from those that determine the final product. In business analytics, the "vision" of the final product is not typically known, and quality is gauged on the insight derived. Quality in business analytics has different dimensions: software quality, model quality, and insights quality. Model quality is determined on accuracy rate, and insights quality is based on creating new actions or improving existing ones. Accuracy rate may vary based on the initial problem. For instance, a 50% accuracy rate may be acceptable if the model addresses a problem never before addressed; thus a MVP version of a model may be acceptable. Acceptable software quality will depend on if the model is a onetime descriptive model or a model to be used in production scoring. Last, at the end of an ASD development cycle, the result is deployed, and the team moves on to the next work item. In ASD, if insight quality is low, the business analytics team might work backward or skip back to steps in CRISP-DM to rework the model.

The challenges of ASD application to business analytics have a common root cause – uncertainty of how and when insights are considered valuable. A basic conflict exists between the experimental nature of business analytics and ASD. Iterative experiments produce learning where one experiment determines the next; there is no understood stopping point or final product.

Conclusion

Traditional project delivery methodologies are not robust enough to address the challenges of big data and or business analytics (Davenport, 2013; Demirkan & Dal, 2014; Gartner Research, 2015). Synthesizing ASD principles, CRISP-DM phases, with emerging best practices to create the business analytics, project delivery framework is a start in evolving business analytics project delivery. ASD values and principles can be leveraged to improve business analytics project delivery, resulting in flexible processes, improved communication, and reduction of rework as outlined by the application of the business analytics project delivery framework results. Some of the challenges encountered applying ASD to business analytics projects include the time-box development cycles and the focus on producing high-quality software. Business analytics value lies in producing new insight, thus producing high-quality software become less of a priority.

References

Abbasi, A., & Adjeroh, D. (2014). Social media analytics for smart health. *IEEE Intelligent Systems, 29*(2), 60–64.

Abbasi, A., Sarker, S., & Chiang, R. H. L. (2016). Data research in information systems: Toward an inclusive research agenda. *Journal of the Association for Information Systems, 17*(2), i–xxxii.

Alnoukari, M. (2015). *ASD-BI: An agile methodology for effective integration of data mining in business intelligence systems*. Hershey, PA: IGI Publishing.

Aston, B. (2017, Feburary 3). 9 project management methodologies made simple: The complete guide for project managers. *DPM*. Retrieved from http://www.thedigitalprojectmanager.com/project-management-methodologies-made-simple/#agile

Bole, U., Popovič, A., Žabkar, J., Papa, G., & Jaklič, J. (2015). A case analysis of embryonic data mining success. *International Journal of Information Management, 35*(2), 253–259.

Beck, K., Beedle, M., Van Bennekum, A., Cockburn, A., Cunningham, W., Fowler, M., et al. (2001). Manifesto for agile software development. Retrieved from http://agilemanifesto.org

Davenport, T. H. (2013). Analytics 3.0. *Harvard Business Review, 91*(12), 64–72.

Davenport, T. (2015). 5 essential principles for understanding analytics. *Harvard Business Review* (hbr.org/2015/10/5-essential-principles-for-understanding-analytics).

Demirkan, H., & Dal, B. (2014, August). The data economy: Why do so many analytics projects fail? *Analytics*. Retrieved from http://analytics-magazine.org/the-data-economy-why-do-so-many-analytics-projects-fail/

Gartner. (2015). Gartner says business intelligence and analytics leaders must focus on mindsets and culture to kick start advanced analytics. *Gartner Newsroom*. Retrieved from https://www.gartner.com/newsroom/id/3130017

Grimes, S. (2006). In BI deployments, methodology does matter. *Intelligent Enterprise - San Mateo, 9*(11), 9.

Halper, F. (2015, January 13). Next-generation analytics and platforms for business success. TDWI Research & Resources. https://tdwi.org/webcasts/2015/01/next-generation-analytics-and-platforms-for-business-success.aspx

Halper, F. (2017, January 11). Data science and big data: Enterprise paths to success. *TDWI Research & Resources*. https://tdwi.org/webcasts/2016/12/big-data-and-data-science-enterprise-paths-to-success.aspx

Jurney, R. (2017). *Agile Data Science 2.0*. Sebastopol, CA: O'Reilly Media.

Keith, E. R. (2006). Agile software development processes a different approach to software design. Retrieved from https://cs.nyu.edu/courses/spring03/V22.0474-001/lectures/agile/AgileDevelopmentDifferentApproach.pdf

Larson, D., & Chang, V. (2016). A review and future direction of agile, business intelligence, analytics and data science. *International Journal of Information Management, 36*(5), 700–710.

Layton, M. C., & Ostermiller, S. J. (2017). *Agile project management for dummies*. Hoboken, NJ: Wiley.

Marbán, Ó., Mariscal, G., & Segovia, J. (2009). A data mining & knowledge discovery process model. In *Data Mining and Knowledge Discovery in Real Life Applications*. Julio Ponce and Adem Karahoca, ISBN 978-3-902613-53-0, pp. 438, February 2009, I-Tech, Vienna, Austria

Piatetsky, G. (2014). CRISP-DM, still the top methodology for analytics, data mining, or data science projects. KDnuggets. Retrieved from https://www.kdnuggets.com/2014/10/crisp-dm-top-methodology-analytics-data-mining-data-science-projects.html

Sahu, A. K. (2016). The criticism of data mining applications and methodologies. *International Journal of Advanced Research in Computer Science, 7*(1), 52–55.

Sim, J. (2014). Consolidation of success factors in data mining projects. *GSTF Journal on Computing (JoC), 4*(1), 66.

Chapter 8
Aligning Operational Benefits of Big Data Analytics and Organizational Culture at WellSpan Health

Gloria Phillips-Wren and Sueanne McKniff

Abstract Our goal in this chapter is to demonstrate the operational benefits that can be gained by implementing real-time, big data analytics in a healthcare setting and the concomitant influence of organizational culture on adoption of the technology. Benefits include improving the quality and accuracy of clinical decisions, processing health records efficiently, streamlining workflow, and improving patient satisfaction. We demonstrate these benefits by investigating patient-physician interactions in a large medical practice at WellSpan Health, and we compare the observed workflow with a modified one made possible with a big data, real-time analytics platform. By comparing these two states, we illuminate the lost opportunity and the value left on the table by legacy behaviors and processes. In addition, we uncover organizational characteristics that create a climate for cultural modification and initial acceptance of big data, real-time analytics in a change-resistant organization. The combination of academic research and practitioner implementation shows that optimization of clinical operations is a key first step toward gaining user acceptance of big data technologies.

Keywords Big data · Analytics · Culture · Healthcare IT · Clinical decision support · Electronic medical record

G. Phillips-Wren (✉)
Loyola University Maryland, Baltimore, MD, USA
e-mail: gwren@loyola.edu

S. McKniff
WellSpan Health, Harrisburg, PA, USA
e-mail: smckniff@wellspan.org

M. Anandarajan, T. D. Harrison (eds.), *Aligning Business Strategies and Analytics*, Advances in Analytics and Data Science 1,
https://doi.org/10.1007/978-3-319-93299-6_8

Introduction

Healthcare in the United States is undergoing a rapid transformation toward using data and analytics to provide improved patient care, evidence-based management, and outcome accountability (Kayyali, Knott, & Van Kuiken, 2013; Wang & Hajli, 2016). Data analytics is based on integrating information from payers, hospitals, laboratories, and physician offices, among others, as fundamental to this shift (Paradise et al., 2013). Such large datasets and the concomitant integration of multiple data types are often referred to as "big data." Analytics associated with big data enables assessment of performance, cost, and utilization metrics and is essential to driving improvements in care management (Paradise, Gold, & Wang, 2013). The next wave of innovation in healthcare is expected to be in real-time health systems based on these data and "the transformation of the healthcare delivery organization into one that is more aware, collaborative and patient-centric" (Runyon, 2016). WellSpan Health, the case study for this paper, is an early adopter of healthcare analytics and is moving rapidly toward enabling and acquiring the underlying data for real-time systems. In this paper, we focus on process analysis and the illumination of workflow inefficiencies to guide the culture change necessary for successful adoption of a new electronic health record (EHR).

The objectives of this chapter are (1) to present management challenges in aligning business strategies and analytics in a complex healthcare organization, (2) to demonstrate operational benefits that can potentially be gained with real-time analytics, and (3) to suggest industry–academic collaboration to address alignment gaps.

WellSpan Health

WellSpan Health is a community-based integrated health system located in South Central Pennsylvania with more than 15,000 employees and 140 patient care locations (WellSpan, 2017). This not-for-profit organization includes a multispecialty medical group of more than 1200 physicians and advanced practice clinicians, a regional behavioral health organization with services for children and adults, and 6 hospitals. WellSpan's hospitals include WellSpan York Hospital (a 580-bed community teaching hospital and trauma center with 7 residency programs, 5 allied health schools, and associated training programs), WellSpan Gettysburg Hospital (a 76-bed acute care hospital), WellSpan Good Samaritan Hospital (a 170-bed acute care hospital), WellSpan Ephrata Community Hospital (a 130-bed acute care hospital), WellSpan Philhaven (a behavioral health hospital), and the 70-bed WellSpan Surgery and Rehabilitation Hospital. WellSpan York Hospital's Regional Resource/Level 1 Trauma Center is accredited by the Pennsylvania Trauma Systems Foundation and is the only accredited trauma center in York, Adams, and Franklin counties.

With a strong focus on the health of the community, WellSpan has been working to transform the delivery of care to meet the changing needs of its central Pennsylvania communities. For example, WellSpan has attempted to reduce hospital medication errors through double bar-code scanning and decision support technologies (Phillips-Wren & McKniff, 2012, 2015). As the healthcare organization has grown with the addition of new hospitals, physicians, and services, WellSpan has increasingly focused its efforts on helping individuals develop a relationship with a primary care physician who can partner with them and the health system to become healthy and stay that way (WellSpan, 2017). Each patient's information is available to providers at any WellSpan location through a sophisticated EMR, and their care is coordinated by teams of physicians, health coaches, social workers, and other professionals.

Although adoption of these technologies offers many benefits, including operational efficiencies, physicians have been slow to adapt to the needed changes in workflow. In the following section, we provide background on big data analytics in healthcare and aligning organizational culture with adoption of the technology.

Background

"Big Data" Analytics

The healthcare industry is generating "big data" in the form of individual patient history due to the large uptake of electronic health records (EHRs). In 2015, 96% of nonfederal acute care hospitals reported use of certified electronic health record (EHR) technology with similar usage at state levels (Henry, Pylypchuk, Searcy, & Patel, 2016). Functionality is also increasing, and efforts are now shifting to interoperability of health information and using technology to support reform of care delivery (Henry et al., 2016). These EHRs contain quantitative data such as laboratory tests, qualitative data such as physician observations, and transactional data such as healthcare delivery records. Although the healthcare industry has utilized scientific inquiry and rigorous analysis of experimental data such as randomized trials to inform practice, big data and associated analytics such as machine learning offer new ways to improve the quality and efficiency of healthcare delivery (Murdoch & Detsky, 2013).

Big data add new dimensions to analytics (Phillips-Wren et al., 2015). Big data can be described as data that have one or more of the characteristics of volume, velocity, and variety (Goes, 2014; SAS 2017), or the three V's. In addition, these data can be generally represented as either structured or unstructured (Agarwal & Dhar, 2014). Volume indicates the huge and growing amount of data being generated, with more data often at higher granularity in EHRs. Velocity indicates the speed at which data are being generated from digital sources such as patient monitoring with wearable sensors, offering the potential for real-time analysis and response. Variety refers to the variation in types of data such as physician observations or X-ray imaging. Structured data reside in spreadsheets and relational databases that impose a structure for storage and access. Semi-structured data lack a strict and rigid format but have identifiable features such as images being tagged with type, date, and patient. Human language is unstructured data of growing importance to analytics using tools to perform activities such as text mining. Other characteristics are sometimes added to the three V's such as variability and complexity (SAS, 2017) or value and veracity (Raghupathi & Raghupathi, 2014) to recognize additional difficulties that organizations encounter in implementing data-intensive applications.

Business intelligence and big data analytics refer to a set of analytical techniques that have been developed to obtain insights from large, complex datasets of varying types using advanced data storage, management, analysis, and visualization technologies (Chen, Chiang, & Storey, 2012). These approaches are based on mathematical models, statistical techniques, decision support methods, data science approaches, and computer science algorithms such as machine learning integrated to address the unique challenges in big data. Generally, analytics can be descriptive, predictive, or prescriptive. Descriptive analytics refers to a description of data and may use exploratory methods to attempt to understand data. Predictive analytics utilizes historical data to predict or forecast a future state. Prescriptive analytics is an emerging field that attempts to find the optimal course of action by examining various possibilities and decision options (Phillips-Wren, Iyer, Kulkarni, & Ariyachandra, 2015). Although many of these methodologies have been standard in data analysis for a long time, in the case of big data, there is a larger amount and variety of data under consideration, as well as the possibility of real-time data acquisition and analysis.

Specific to healthcare and based on analysis of 26 big data implementation cases, derived benefits from analytics were classified into five benefit categories: IT infrastructure, operational, organizational, managerial, and strategic (Wang, Kung, & Byrd, 2016). The two benefits noted most often were IT infrastructure (reduce system redundancy, avoid unnecessary IT costs, and transfer data quickly among healthcare IT systems) and operational benefits (improve the quality and accuracy of clinical decisions, process a large number of health records in seconds, and reduce the time of patient travel) (Wang et al., 2016). "This implies that big data analytics has a twofold potential as it implements in an organization. It not only improves IT effectiveness and efficiency, but also supports the optimization of clinical operations. In addition, [the] results also indicate that big data analytics is still at an early stage of development in healthcare due to the limited benefits of big data analytics at the organizational, managerial, and strategic levels" (Wang et al., 2016, p. 8).

In this paper, we focus on the optimization of clinical operations by investigating one service delivery area in WellSpan Health that can use big data analytics to streamline operations, improve patient safety, and enhance clinical decision-making. In this case, there is potential to impact all three levels of the system.

"Big Data" in Healthcare Practice

The healthcare industry appears to be repeating the trend of computerization and data management that has occurred in other industries (Sanders, Burton, & Protti, 2013). Sanders et al. (2013) describe three phases. Phase I consists of data collection and transaction-based processing and is reflected in EMR adoption. Phase II is information sharing, facilitated in healthcare by Healthcare Information Exchanges (HEI). Phase III is the data analysis phase characterized by enterprise data warehouses and analysis of "small" and "big" data. "Small" data are not necessarily small in volume – they may be huge datasets; they are simply able to be collected internally and analyzed using existing tools. In many organizations, analytics associated with even "small" data represents a significant step forward.

Moving to "big data" analytics presents even more challenges in healthcare settings. One way to overcome the challenges is to follow a generalized methodology consisting of four steps: (1) a concept statement to establish the need based on the three V's; (2) a proposal development stage; (3) fleshing out of the methodology including data and platform identification, data acquisition and cleaning, data transformation, and data analysis; and (4) deployment including testing, evaluation, and validation (Raghupathi & Raghupathi, 2014). "This process differs from routine analytics only in that the techniques are scaled up to large data sets" (Raghupathi & Raghupathi, 2014, p. 7). As we will see later, the specific methodology utilized by WellSpan Health follows this general process.

A more specific Healthcare Analytics Adoption Model is shown in Fig. 8.1 and provides a way to assess expansion of analytics capabilities and maturity in data sources, complexity, data literacy, and data timeliness (Sanders et al., 2013). To progress through the steps, an organization must have standard ways of collecting data and assessing its validity, integrating data, automating reporting, reducing variability in processes, tailoring patient care, predicting outcomes to suggest interventions, and tailoring patient care. The nine levels shown in Fig. 8.1 move from inconsistent versions of the truth at Level 0 through standardization processes in

Fig. 8.1 Healthcare analytics adoption model. (Adapted from Sanders et al., 2013)

Level 8: Personalized prescriptive patient care
Level 7: Predictive risk models for patient intervention
Level 6: Patient care guided by population metrics
Level 5: Reduced variability in care processes
Level 4: Efficient, consistent, adaptable external reporting
Level 3: Efficient, consistent, available internal reporting
Level 2: Standardized vocabulary and patient registries
Level 1: Collection and integration of core data
Level 0: Inefficient, inconsistent point solutions

Levels 1 and 2. Internal standardization can be obtained with an enterprise data warehouse that employs consistent language, norms, and governance.

Analytics is employed on Level 3 to develop automated internal reports and new knowledge with internal data for executive decision-making, and it is extended to impact external reporting in Level 4 and clinical best practice in Level 5. Levels 6–8 require data beyond the organization and can be considered big data. Level 6 at the point-of-care uses population metrics to guide patient care, and those metrics are drawn by analysis of the larger world body of patients. Predictive risk models in Level 7 require external data to include in the analysis such as collaboration between physicians, hospitals, payers, and patients. Level 8 utilizes analytics toward patient health optimization and requires personalized data such as genetic data.

Organizations do not move through these phases linearly, and they may be working on multiple levels simultaneously. However, "the return on investment of EMRs … will not be realized … until the healthcare industry … commits culturally to the exploitation of analytics, – that is, to become a data-driven culture, incented economically to support optimal health at the lowest cost" (Sanders et al., 2013, p. 8).

Even with the advantages of these processes, there is resistance in healthcare. Evidence-based medicine (EBM) is "the conscientious, explicit, and judicious use of current best evidence in making decisions about the care of individual patients" (Sackett, Rosenberg, Gray, Haynes, & Richardson, 1996). "EBM stands in contrast to anchoring decisions on personal habits, tangible and intangible incentives unrelated to care, or medical traditions that have little or no empirical validation" (Fichman, Kohli, & Krishnan, 2011). Barriers to widespread adoption of analytics include lack of knowledge and misconceptions about the actual effectiveness of treatment, the difficulty of diffusing that knowledge, and practitioner resistance that is often connected to a desire for autonomy, incentive conflicts, and fear of litigation (Fichman et al., 2011). Many of these issues of technology adoption and culture change have been faced and studied by the information systems community for decades (Karahanna, Straub, & Chervany, 1999), and the lessons learned can be applied to healthcare.

"In healthcare, the foremost concern for management is the people that compose it – the key stakeholders – be they patients, physicians, nurses, and other medical staff, referring providers, or representatives from the local community. Empowering these individuals and increasing the quality and transparency of decision-making are key goals for any business analytics initiative. Therefore, the organization needs to establish business analytics as an organizational and cultural objective, a component of its long-term strategy. However, to realize these benefits, clinicians, support staff, and leadership all need to understand and appreciate the importance of business analytics as tools and as a fundamental process within the organization. Otherwise, the organization will continue to underinvest and staff will be skeptical of the value of recording data as a matter of course" (Ward, Marsolo, & Froehle, 2014, p. 577).

Organizational Culture and Technology Adoption

Culture has been framed in various ways, for example, as ideologies, beliefs, assumptions, shared values, collective will, norms, practices, symbols, rituals, myths, ceremony, and tacit versus explicit components (Leidner & Kayworth, 2006). Organizational culture is an important factor in absorbing and implementing new technology in a healthcare setting (Caccia-Bava, Guimaraes, & Harrington, 2006). A review of the literature shows that there is a range of social, technical, and organizational characteristics that need to be managed to ensure that technology innovations are useful to healthcare organizations and individuals; however, these factors are interrelated and complex (Cresswell & Sheikh, 2013).

Information systems research has a rich history of investigating the influence of culture (e.g., national, organizational, group) and the adoption of IT in organizations (Leidner & Kayworth, 2006). In particular, technology acceptance and use has been studied at length in the information systems community (Phillips-Wren & McKniff, 2015). Perhaps the best-known theoretical model is the technology acceptance model (TAM) (Davis, 1989) and its extensions based on the theory of reasoned action (Fishbein & Ajzen 1975) that attitudes and norms (e.g., influenced by culture) predict behavioral intention which predicts actual behavior. TAM has been utilized to understand physician acceptance of telemedicine, decision support systems in primary care, physicians' intention to accept innovation, adverse event reporting, technology within hospitals, and healthcare professionals' intention to utilize healthcare information systems (Phillips-Wren & McKniff, 2012, 2015). Technology adoption is a necessary important first step toward a data-driven organization, and it is the aspect that we focus on in this paper.

In the following section, we apply these concepts to analyze a situation in a healthcare organization dedicated to moving toward real-time analytics and achieving the benefits of a data-driven culture for its organizational processes and patient care.

Methodology and Discussion of Results

Issues, Controversies, and Problems

As the WellSpan organization began preparations for the substantial undertaking of replacing multiple legacy systems with an enterprise-wide EHR, a variety of departments were identified as being high risk for adoption of the workflows necessary for successful EHR implementation. Direct observation of the service delivery processes within these departments reinforced these concerns. "Workflows must be designed in a way that assures the important data elements will be captured during a visit and that these tasks minimally disrupt workflow, particularly expensive resources such as nurses and physician" (Ward, Marsolo, & Froehle, C., 2014, p. 378).

In response to an invitation from leadership of orthopedic services at WellSpan, and with full disclosure of the intent to evaluate workflow as it pertained to the EHR, we observed various teams during regularly scheduled office hours. Our initial approach provoked a Hawthorne effect where the person being observed changes their behavior in response to observation. Thus, the care teams attempted to use the computer system in ways that further broke their processes and reinforced their beliefs of computer-generated process inefficiencies. On a few occasions, the orthopedic team used this opportunity of having a captive audience to share testimonials and demonstrate functionality flaws. These dialogues introduced additional distractions and delays to such an extreme that the observers chose to remove themselves, so throughput, for the day, might be restored.

To mitigate some of the difficulties with data acquisition, a different approach was introduced in attempt to increase the opportunities of capturing accurate workflow examples. Trusted in-house personnel with clinical workflow experience shadowed the clinical support staff. These individuals gained trust of the care team and provided value in real-time while simultaneously witnessing the work-arounds. More importantly, they gained awareness via uncensored remarks justifying the work-arounds. All management levels, including site directors and practice managers, were actively involved in this undertaking which became known to the process improvement team as "WOW" WellSpan Orthopedic Workgroup.

Physicians shared a common fear that any amount of time spent interacting with the EHR is time not spent with a patient and, thus, would increase patient visit times and ultimately decrease visit volume. This deep-seated cultural belief was the foremost obstruction to the project. A paradigm shift needed to occur. The belief that the EHR is just a slow electronic form of documentation is a perhaps a valid perception, albeit outdated. Value-added data served up just-in-time to provide guidance during the fast-paced delivery of orthopedic ambulatory care needed to be proven.

"Current State" Workflow Observed in Practice

Figure 8.2 shows a traditional cross-functional workflow diagram representing the findings of our observations on the "current state." It shows the role-based activity steps and handoffs during a typical ambulatory office visit for orthopedic evaluation or postoperative visit. Despite having a functioning EHR, physicians and their support teams employed work-arounds designed to decrease the physician's need to interact with the computer. As shown in the diagram, the physician verbally requests the room location of the next patient to be evaluated. The support staff pause, think, and respond verbally. Following the physician's exam, orders are verbally communicated in the charting area hallway. The details necessary to complete the electronic orders frequently require additional clarification. Decision support systems and best practice advisories appear during the order placement, and resolution of the

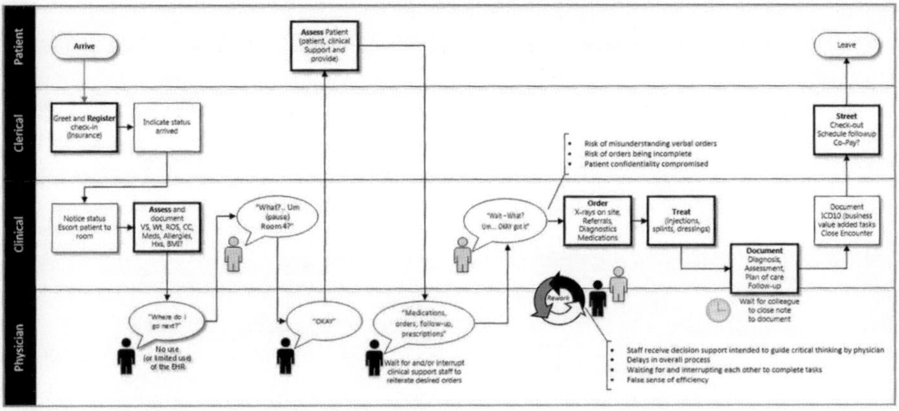

Fig. 8.2 Current state – office throughput

alert often requires the expertise of the physician. This creates a recurring situation of waiting and interrupting one another for task completion or perhaps a riskier practice of the alert being addressed by a non-physician.

Note writing is another area in which the support staff devised a work-around to alleviate some of the documentation burden for the physician. However, only one person can contribute to a note at a time. This created a situation of waiting between colleagues to finish and release the note. One enhancement recommended by the care team during our observations was to allow for multiple simultaneous contributors to a note. While this might seem to make documentation faster, it has obvious inherent safety concerns.

Clinical support staff can be inadvertently put in situations in which they are practicing above their certification/license in the name of physician efficiency. Lack of standard processes and physician preferences introduced process variation that challenges support staff to work effectively with different physicians and contributes to staff attrition and retention issues.

Workflow Process for "Future State"

We suggested that by using big data technologies and analytics with the EHR, the process could be significantly simplified to a "future state" with improvements as shown in Fig. 8.3. In this workflow, the physician uses technology associated with the EHR to locate the patient; capture observations and recommendations during the patient encounter; record heath strategies such as medication, treatment, follow-up, and referrals; and move to the next patient. We suggest that although cultural changes are required to implement the new workflow, the advantages will be significant. The physician's expertise is captured accurately, decisions are documented, patient prior encounter information is available, patient history can be incorporated

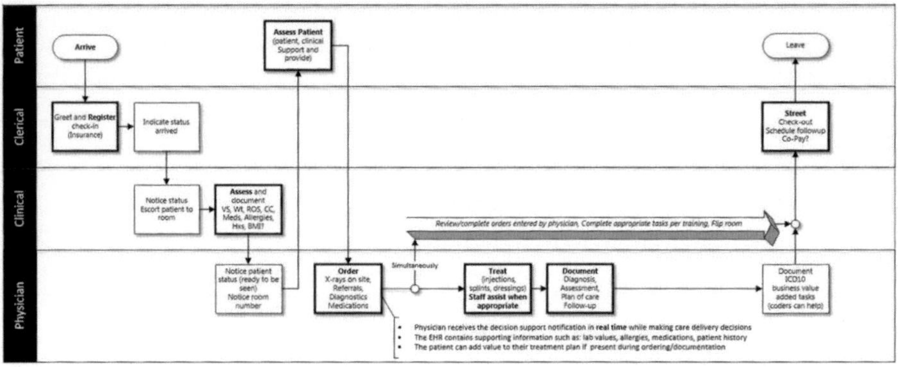

Fig. 8.3 Future state – office throughput

into treatment plans, best practice alerts can guide decision-making in real-time, and clinical staff contact with the patient is more effective. The support staff will become more available to effect office throughput because of the previous opportunity cost of their time waiting and doing nonsupport staff responsibilities. Support staff will not be waiting to clarify verbal orders and other rework that was out of the scope of their certification/license. In addition, the capture of structured and unstructured data can be incorporated into the EHR and used to inform real-time decision-making and reporting.

Technology Acceptance and Cultural Challenges

We used the traditional tool of a cross-functional diagram to illuminate what we thought were obvious risks and delays. We created a "current state" diagram (Fig. 8.2) based on our observations and a proposed "future state" diagram (Fig. 8.3). We presented these documents to a few of the orthopedic physicians. Our technique failed. We did not successfully persuade them that a workflow change was necessary for their efficiency or for their readiness to adapt to the impending implementation of a new EHR. The culture is so ingrained, along with the fact that diagrams were developed by non-orthopedics personnel, that the diagrams were considered an inaccurate representation of the current state. Physicians feared that the recommended process changes would be detrimental to their office efficiency and, subsequently, their livelihoods.

A different approach to overcome organizational resistance to technology was needed. After discussion with leaders, we presented the workflows at an orthopedic all-staff and all-provider meeting and asked them for help validating our findings. We used custom animation via PowerPoint to make the workflow advance one activity step at a time. We replaced some of the words and squares with clip art representing physicians and staff. We did this slowly and void of judgment. The energy in the room became palpable. The staff clapped and cheered as group consensus grew and the gaps and rework were illuminated. The interactive approach overcame initial resistance as the group participated in developing the process flow.

The approach used during the team meeting was a catalyst of multiple downstream efforts to change culture and affect efficiency. The orthopedic care team realized that ineffective processes had accidently evolved in the name of efficiency. The phrase "efficiency theater" was a light way for staff to refer to a situation where rework and reduced waiting time resulted when colleagues assisted each other in using the technology effectively.

Discussion of Results: Digital Transformation at WellSpan

Within WellSpan Health, an organizational goal is the use of technology to improve the delivery of patient care. WellSpan's mission is "Working as one to improve health through exceptional care for all, lifelong wellness and healthy communities." The executive leadership team at WellSpan Health has embraced the challenge to provide tools necessary to deliver exceptional care. A reliable EHR, accurate real-time analytics, and clinically led information technology initiatives have provided a foundation for success.

However, this technology adoption project was initially interpreted as risky while not adding value. Using technology was viewed as slowing down the number of patients that could be seen in a day, and time is critical since charges are accrued per patient. In addition, healthcare professionals prefer to interact with humans rather than a computer. They are trained to assess a situation and make decisions without technology, so introducing technology into the workflow was not clearly an advantage. Yet they rely on discrete patient-specific data and situational health information when applying their knowledge of medicine to safely formulate their care delivery decisions. The EHR will be considered a value-added tool when physicians and nurses experience value in real-time through data analytics such as decision support alerts that interrupt the ordering of an inappropriate medication, or a population health alert that recommends an appropriate diagnostic screening exam, or a timesaving alert indicating that a procedure will not be covered by the patient's insurer. When providers experience the embeddedness of the EHR in the delivery of safe healthcare, then digital transformation will begin to occur.

Clinical stakeholders must be involved in all phases of technology development and implementation in order to incentivize physicians and clinical end users to adopt and embrace it. The collective experience of interacting with the EHR defines usability. Issues such as slow log-on times, number of clicks to perform a task, and nonintuitive design contribute to resistance and subsequently underutilization by the end user. Just-in-time support and at-the-elbow guidance may mitigate the steep learning curve that historically has contributed to the delegation of EHR responsibilities to residents and junior staff. Such delegation further delays the hands-on experience required for adoption and increases patient risk potential because decision support is being addressed by less experienced providers.

Another technique to positively affect the rate of digital transformation is to listen to the physicians' concerns and mitigate them when possible. If the physician is incentivized for visit volume, they must be provided tools that do not impede their ability to function at an optimal capacity. In our case, it was important not to replicate bad processes or simply automate current paper records. Lean methodologies should be applied to the workflows surrounding the physician's workload such as strategizing and standardizing tasking, formally delegating prescription refill responsibilities, or alleviating time burdens via upstream data collection such as documentation of the patient's preferred pharmacy.

By clearly communicating organizational goals and subsequent anticipated results, members of an organization can collaborate to achieve a common goal. One example is to achieve a specific level/stage of EMR adoption model. In 2015, WellSpan Health's hospitals were judged by a US organization, Healthcare Information and Management Systems Society (HIMSS), an American not-for-profit organization dedicated to improving healthcare in quality, safety, cost-effectiveness, and access through the best use of information technology and management systems. WellSpan received two Electronic Medical Record Adoption Model (EMRAM) certificates of achievements from HIMSS. Figure 8.4 shows the percent of hospitals achieving that level in 2015. WellSpan Surgery and Rehabilitation Hospital in York, Pennsylvania, achieved HIMSS EMRAM Level 7, and WellSpan Good Samaritan Hospital in Lebanon, Pennsylvania, achieved HIMSS EMRAM Level 6 designation in 2015.

The current HIMSS EMRAM is shown in Fig. 8.5 along with the percent of hospitals achieving each level in 2017. It differs somewhat from the Generalized Healthcare Analytics Adoption Model discussed earlier (Fig. 8.1) and the HIMSS EMRAM 2015 (Fig. 8.4). HIMSS EMRAM 2017 has seven stages and does not currently require the use of personalized data such as genetic markers as outlined in Level 8 of the Sanders et al.'s (2013) model. We discussed the generalized model in Fig. 8.1 as aspirational and applicable to all healthcare organizations, national and international, while the EMRAM 2017 model in Fig. 8.5 is an instance of a metric-driven evaluative model developed by a specific US organization (HIMSS, 2017). These standards continue to evolve.

In 2015, at the time WellSpan Surgery and Rehabilitation Hospital achieved stage 7, only 4.2% of the hospitals in the United States had achieved this designation, while 27.1% had achieved stage 6 with WellSpan Good Samaritan Hospital. In 2017 Q3, the percentage of US hospitals achieving stage 7 had risen to 6.1%, while the percentage in stage 6 was 32.7% (HIMSS, 2017). Thus, IT adoption is improving in US hospitals, although more slowly than proponents desire (Keller et al., 2017).

Stage	Cumulative Capabilities 2015	2015 Q4 WellSpan awarded Level 7
7	Complete Electronic Medical Record (EMR); continuity of care transactions to share data; data warehousing; data continuity with emergency department; ambulatory; out-patients	4.2%
6	Physician documentation with structured templates; full clinical decision support system with variance and compliance; full images	27.1%
5	Closed-loop medication administration	35.9%
4	Computerized physician order entry (CPOE); clinical decision support with clinical protocols	10.1%
3	Nursing/clinical documentation (flowsheets); clinical decision support system with error checking; images available outside Radiology	16.4%
2	Clinical data repository (CDR); controlled medical vocabulary; clinical decision support; may have document imaging; Health Information Exchange (HIE) capable	2.6%
1	Ancillaries – Laboratory; Pharmacy; and Radiology/Cardiology - all installed	1.7%
0	All three ancillaries not installed	2.1%

Fig. 8.4 United States Electronic Medical Record (EMR) adoption model of HIMSS analytics 2015

Bridging the Gap Between Research and Practice

A major issue in the implementation and effective use of big data analytics in an organization is convincing practitioners that the required changes are worth the effort. One way that the academic community can contribute to overcoming this perception is by theoretically researching benefits of big data analytics and documenting those benefits with real use cases.

In healthcare, academics can provide evidence that a larger population of data acquired by the EHR will improve evidence-based decision-making and lead to improved patient care. These new data enhance generalizability of medical decisions by including demographics and patient characteristics that cannot be captured

Stage	Cumulative Capabilities 2017	2017 Q3
7	Complete Electronic Medical Record (EMR): Includes External Health Information Exchange (HIE); data analytics; governance; disaster recovery; privacy and security	6.1%
6	Technology-enabled medication, blood products, and human milk administration including risk reporting	32.7%
5	Documentation by physician using structured templates with intrusion/device protection	33.5%
4	Computerized physician order entry (CPOE) with clinical decision support; nursing and allied health documentation with basic business continuity	10.1%
3	Nursing and allied health documentation; electronic medication administration records; role-based security	12.6%
2	Clinical data repository (CDR) including internal interoperability and basic security	1.9%
1	Ancillaries – Laboratory; Pharmacy; and Radiology/Cardiology information systems with image management system	1.5%
0	All three ancillaries not installed	1.6%

Fig. 8.5 United States Electronic Medical Record (EMR) adoption model of HIMSS analytics

in randomized trials that are necessarily limited by practicality. To be most effective, big data analytics should be delivered into the hands of the most experienced professional to affect decision-making, so the physician should personally see the data, the alerts, and best practices suggested by the data.

Academics can also apply process-flow methodologies to enhance operations using big data analytics and operational research techniques. The concept of having additional human resources to relieve physicians from tasks that do not require their expertise is logical at first glance. Physicians are trained professionals who should not be performing tasks such as record keeping. The authors agree with this premise when the task was simply transcription of notes into electronic format. However, the increasing capacity of computers to assist in complex human decision-making creates new opportunities. Using big data analytics such as that made possible with EMRs, computers can correlate patient outcomes with healthcare, identify trends and best practices in the larger population, provide individualized recommendations and alerts, and with greater precision provide information to mitigate patient harm. Thus, physicians need to interact directly with the technology as records are being created.

In the classroom, case studies such as WellSpan serve to illustrate the applicability of analytics methods to the real world and, especially for graduate students, provide an analogy to a student's own organization. It is especially powerful to have business partners as speakers to instructional groups. This two-way dialogue between educators and members of the business community lends credence to pedagogy and relevance to the concomitant research.

Conclusion

This paper reports research on the environmental conditions needed to obtain value from big data analytics in a healthcare setting. We focused on an orthopedics unit at WellSpan Health and observed the workflow and the interaction of physician, staff, patient, and documentation processes. We found that simply having an electronic medical record available was not enough. Although real-time analytics has the potential to assist healthcare professionals with decision-making, underlying cultural processes can interfere with effective use of available data. In our case, optimization of clinical operations was a key first step toward utilizing electronic medical record systems effectively to improve patient safety and enhance clinical decision-making.

Big data analytics will continue to grow in the healthcare sector due to its proven ability to enhance medical decision-making and improve operational efficiency. Physicians and healthcare providers are becoming increasingly trusting and subsequently reliant on real-time, big data during the decision-making phases of the delivery of patient care. Not just the use of technology but the use of smart timely data will continue to be further embedded into the workflow of personnel caring for patients.

Acknowledgments We extend special thanks to Mark A Deitch, MD, MBA, Vice President, Orthopedic Service Line, WellSpan Health, for his guidance and support.

References

Agarwal, R., & Dhar, V. (2014). Editorial: Big data, data science, and analytics: The opportunity and challenge for IS Resesarch. *Information Systems Research, 25*(3), 443–448.

Caccia-Bava, M., Guimaraes, T., & Harrington, S. (2006). Hospital organization culture, capacity to innovate and success in technology adoption. *Journal of Health Organization and Management, 20*(3), 194–217.

Chen, H., Chiang, R., & Storey, V. (2012). Business intelligence and analytics: From big data to big impact. *MIS Quarterly, 36*(4), 1165–1189.

Cresswell, K., & Sheikh, A. (2013). Organizational issues in the implementation and adoption of health information technology innovations: An interpretative review. *International Journal of Medical Informatics, 82*, e73-e86.

Davis, F. D. (1989). Perceived usefulness, perceived ease of use, and user acceptance of information technology. *MIS Quarterly, 13*(3), 319–339.

Fichman, R., Kohli, R., & Krishnan, R. (2011). The role of information systems in healthcare: Current research and future trends. *Information Systems Research, 22*(3), 419–428.

Fishbein, M., & Ajzen, I. (1975). *Belief, attitude, intention, and behavior: An introduction to theory and research*. Reading, MA: Addison-Wesley Publishing Co.

Goes, P. (2014). Editor's comments: Big data and IS research. *MIS Quarterly, 38*(3), iii–viii.

Henry, J., Pylypchuk, Y., Searcy, T., & Patel, V. (2016, May). Adoption of electronic health record systems among U.S. Non-Federal Acute Care Hospitals: 2008–2015. *The Office of National Coordinator for Health Information Technology.* Retrieved from https://dashboard.healthit.gov/evaluations/data-briefs/non-federal-acute-care-hospital-ehr-adoption-2008-2015.php

HIMSS Analytics. (2017). Electronic medical record adoption model. Retrieved from http://www.himssanalytics.org/emram

Karahanna, E., Straub, D., & Chervany, N. (1999). Information technology adoption across time: A cross-sectional comparison of pre-adoption and post-adoption beliefs. *MIS Quarterly, 23*(2), 183–213.

Kayyali, B., Knott, D., & Van Kuiken, S. (2013). *The big-data revolution in US health care: Accelerating value and innovation* (pp. 1–13, vol. 2(8)). Mc Kinsey & Company. Retrieved from http://www.mckinsey.com/industries/healthcare-systems-and-services/our-insights/the-big-data-revolution-in-us-health-care

Keller, E., Crowley-Matoka, M., Collins, J., Chrisman, H., Milad, M., & Vogelzang, R. (2017). Fostering better policy adoption and inter-disciplinary communication in healthcare: A qualitative analysis of practicing physicians' common interests. *PLoS One, 12*(2), e0172865.

Leidner, D., & Kayworth, T. (2006). A review of culture in information systems research: Toward a theory of information technology culture conflict. *MIS Quarterly, 30*(2), 357–399.

Murdoch, T., & Detsky, A. (2013). The inevitable application of big data to health care. *Journal of the American Medical Association, 309*(13), 1351–1352. https://doi.org/10.1001/jama.2013.393

Paradise, J., Gold, M., & Wang, W. (2013, October 01). *Data analytics in Medicaid: Spotlight on Colorado's accountable care collaborative*. The Kaiser Foundation. Retrieved from http://kff.org/medicaid/issue-brief/data-analytics-in-medicaid-spotlight-on-colorados-accountable-care-collaborative/

Phillips-Wren, G., Iyer, L. S., Kulkarni, U., & Ariyachandra, T. (2015). Business analytics in the context of big data: A roadmap for research. *Communications of the Association for Information Systems, 37*, 23.

Phillips-Wren, G., & McKniff, S. (2012). Fusing decision support into the fabric of healthcare to prevent medication errors. In *DSS* (pp. 27–36).

Phillips-Wren, G., & McKniff, S. (2015). Beyond technology adoption: An embeddedness approach to reduce medication errors. *Journal of Organizational Computing and Electronic Commerce, 25*(2), 1–20.

Raghupathi, W., & Raghupathi, V. (2014). Big data analytics in healthcare: Promise and potential. *Health Information Science and Systems, 2*(1), 3.

Runyon, B. (2016, May 11). *Industry vision: The real-time health system* (Gartner ID: G00308258.). Gartner.

Sackett, D. L., Rosenberg, W., Gray, J., Haynes, R., & Richardson, W. (1996). Evidence based medicine: What it is and what it isn't. *British Management Journal., 312*(7023), 71–72.

Sanders, D., Burton, D., & Protti, D. (2013). The healthcare analytics adoption model: A framework and roadmap. HealthCatalyst. Retrieved from http://healthsystemcio.com/whitepapers/HC_analytics_adoption.pdf

SAS. (2017). *Big data – What is it and why it matters.* Retrieved from https://www.sas.com/en_us/insights/big-data/what-is-big-data.html

Wang, Y., & Hajli, N. (2016). Exploring the path to big data analytics success in healthcare. *Journal of Business Research, 70*, 287–299. Retrieved from https://doi.org/10.1016/j.jbusres.2016.08.002.

Wang, Y., Kung, L., & Byrd, T. (2016). Big data analytics: Understanding its capabilities and potential benefits for healthcare organizations. *Technological Forecasting and Social Change* Retrieved from https://doi.org/10.1016/j.techfore.2015.12.019

Ward, M., Marsolo, K., & Froehle, C. (2014). Applications of business analytics in healthcare. *Business Horizons, 57*(5), 571–582.

WellSpan. (2017). Retrieved from https://www.wellspan.org/

Chapter 9
HR Analytics: Human Capital Return on Investment, Productivity, and Profit Sensitivity: A Case of Courtyard Marriott Newark at the University of Delaware

Ali A. Poorani and William A. Sullivan

Abstract The objective of this case study is to apply human capital analytics, more specifically, human capital return on investment, human resources productivity, and compensation efficiency at the Newark Courtyard Marriott Hotel, University of Delaware, and investigate if such analytics adds new outlooks beyond the usual metrics used by lodging enterprises. The study presents quantitative metrics and trend analysis for a 3-year period at this business unit. In addition, the case study provides measures that help management to identify and address inefficiencies, as well as the productivity of its human capital. The study also highlights the benefits of *Bridging Practice and Theory*.

Keywords Human capital analytics · Productivity of human resources · Profit sensitivity · Big data · Case study

Introduction

Obviously, a strategy of maintaining a culture of employee engagement and retention requires investment in human capital, but little is known about the financial results of investment in human capital (Marler & Boudreau, 2017). Both executives at the corporate level and managers at the property level need to justify the cost-benefit of investment in human capital by leveraging the power of

A. A. Poorani (✉)
Hospitality Associates for Research & Training (HART), Hospitality Business Management, Alfred Lerner College of Business and Economics, University of Delaware, Newark, DE, USA
e-mail: poorani@udel.edu

W. A. Sullivan
Courtyard Newark at the University of Delaware, Newark, DE, USA
e-mail: bsullivan@shanercorp.com

M. Anandarajan, T. D. Harrison (eds.), *Aligning Business Strategies and Analytics*, Advances in Analytics and Data Science 1,
https://doi.org/10.1007/978-3-319-93299-6_9

real data that not only supports a culture of engagement but also adds to financial results. As cited by Stevens (2015), most of the narratives when addressing HR issues to board members and executives are derived from expertise, intuition, and gut feelings rather than statistical analysis of the data. As Stevens (2015) quoted Brad Church, HR business partner at the Williams Cos. Inc., "most of the people we were talking to wanted something that could carry more weight than just my best judgment."

A study of big data across a variety of industries (Erikson & Rothberg, 2015) revealed that gaming and hotels earned about average scores on knowledge management (KM) but relatively low scores in utilizing big data and competitive intelligence (CI). This is surprising, since hotel and gaming companies possess large volumes of data from loyalty programs, customer data, and service operations. Perhaps, this deficiency could be attributed to the shortage of skilled employees with the knowledge to utilize big data and lack of widespread learning from other industries. A study of 328 business executives by TDWA, Best Practices Report (Halper, 2015), revealed that more than 52% of executives indicated that *the lack of skilled personnel* was the number one challenge for next-generation analytics.

The objective of this case study is to apply human capital analytics, more specifically, financial return on human capital investment at the Newark Courtyard Marriott Hotel, and investigate if such analysis adds new outlooks beyond the usual metrics used by lodging enterprises. The study presents quantitative metrics and a standardized financial approach to investment in human capital at this business unit. In addition, the case study provides measures that management can identify to address inefficiencies as well as the productivity of its human capital. The study also highlights the benefits of *Bridging Practice and Theory.*

Background

As compared to many other businesses, human capital costs in the hotel industry are highly significant. According to the *Lodging Magazine*, posted by Robert Mandelbaum (2014), "In 2013, 44.8 percent of all dollars spent to operate a hotel in the United States went to pay for labor-related costs, making it the single largest expense item for operators." And, overall the labor costs ranged from roughly 35% at limited-service and extended-stay properties to 48% at convention hotels and resorts for the same year. Among the contributing factors for high human capital costs are high turnover rates ranging from 25% for managerial positions to about 60% for line-level employees with a total turnover cost per employee ranging from $2,604 to 14,019 (Tracey & Hinkin, 2006).

Kline and Harris (2008) attributed the failure of hoteliers to expect accountability for the investment into employee development and return on investment (ROI) in training to hard-to-identify human capital bottom-line contributions, time constrains, lack of reliable information, and reliable tracking methods for employee contributions. Today, big data and related analytics are revolutionizing the way we see and process the world. According to Mayer-Schonberger and Cukier (2013) and McMillan (2016), the use of analytics and big data suggests implementing a system of accountability, including measuring success with analytics and big data, and delivering value that will be credible to top executives.

Big Data a Definition

Big data is distinguished from regular data by (the three Vs) volume, velocity, and variety (Chintagunta, Hanssens, & Hauser, 2016) defined as follows:

- Volume— massive size.
- Velocity— how fast data are changing or coming in.
- Variety— multiple formats, such as Facebook, Tweeter, encrypted packages, etc.

Brands (2014) expanded on the 3Vs to include:

- Value—usefulness
- Venue—location
- Vocabulary—context

For example, Walmart collects approximately 2.5 petabytes (1 petabytes = 1,000,000 gigabytes) of information per hour including information about transactions, customer behavior, location, and other devices (Bradlow, Gangwar, Kopalle, & Voleti, 2017). The challenge for organizations is how to leverage this massive information and add value to their organizations. Business intelligence (BI) offers a broad range of analytical tools to access, mine, and analyze big data.

Revolution of HR Analytics

Although measurement of human resources has been around since early 1900s, Marler and Boudreau (2017) conducted an evidence-based literature review of published peer-reviewed articles that addressed HR analytics. Their analysis showed that the volume of research regarding modern HR analytics in high-quality peer-reviewed journals is limited. Using an innovation adoption method, their study identified 60 articles on this topic of which only 14 articles were published in quality peer-reviewed journals. Additionally, 10 out of the 14 articles did not involve hypothesis testing but were short illustrative case studies. One can conclude that research on HR analytics is still in its infancy, and there is a need for future studies.

Utilizing the published reviewed articles, Marler and Boudreau (2017) offered the following definition for HR analytics:

> A HR practice enabled by information technology that uses descriptive, visual, and statistical analyses of data related to HR processes, human capital, organizational performance, and external economic benchmarks to establish business impact and enable data-driven decision-making.

Bassi (2011) offered a similar definition that HR analytics is an evidence-based approach; it could include an array of simple reporting metrics all the way to predictive modeling. There is no one size fits all. The analytics could take a form of a single environment to global analytics across various environments, such as regulatory, geographic, etc. Beyond the definition, in their study, Marler and Boudreau (2017) focused on the following four questions:

1. How does HR analytics work? In summary, 6 of the 14 articles addressed the LAMP model (logic, analytics, measures, and processes), the key elements necessary to understand the cause-effect relationships in HR analytics.
2. Why does HR analytics work? Evidence-based research regarding this question is very limited. We hope that our case study will shed more light on how and why the HR analytics works in the lodging industry. Aral, Brynjolfsson, and Wu (2012), using a sample of 189 firm-level data collected over 5 years from 1995 to 2006, showed that HR analytics alone did not result in higher productivity, but firms that combined human capital management (HCM) software and pay for performance method achieved significantly more productivity. In this case, aligning incentives and employee behavior (logic), using HCM (analytics) with pay for performance (process), and measures from HCM software neatly corresponds to the LAMP model.
3. What are the outcomes of HR analytics? According to Bassi (2011), the end goal is to improve individual and organizational performance. An empirically tested study by Aral et al. (2012), along with six nonempirical case studies, demonstrated a strong relationship between HR analytics and financial results. A study by Harris, Craig, and Light (2011) offered a divergent view and asserted that cost savings on HR process are unlikely to produce significant financial impact based on the notion that administrative costs typically constitute about 3% of company's sales. In our assessment, although this number might accurately represent many business segments, lodging industry is highly labor-incentive, and this conclusion may not accurately represent this industry. Other less promising studies included a low or moderate use of HR analytics to measure the relationship between HRM process and people and business impact across Fortune 1000 firms (Falletta, 2013) or Fortune 500 companies (Lawler & Boudreaau, 2015) demanding more research and explanations.
4. What moderating factors affect HR analytics outcome? Marler and Boudreau (2017) concluded that HR analytics to be successful required three moderating factors, namely having HR professional analytical skills, acquiring managerial buy-in, and using HR information technologies. This conclusion is further supported by skepticism about the ability of human resources professionals to effec-

tively employ data analytics as well as deliver data-driven and business-oriented organizational results (King, 2016).

Often there is a disconnect between data that HR professionals use to track HR metrics, such as turnover ratios, demographics, promotion rates, etc., and their effects on total human capital return on investment for the unit or the organization as a whole. In this chapter, we work to break this disconnect by using accounting data (real and accurate data) to provide a quantitative measure of human capital cost for the organization and link those costs to the financial success of the firm. As an example, at Best Buy, a 0.1% increase in investment in employee engagement has been measured to create more than $100,000 in the store's annual operating income (Davenport, Harris, & Shapiro, 2010). We propose that such analyses can be carried out at the unit level, as well as the entire organization for the hotel industry. As these methods of data collection build up, we foresee useable trends emerging that can be exploited to measure progress and cost-benefits of the human capital components to the organization's total cost structure.

One controversy regarding HR analytics is who should take the lead, HR or finance? Bassi (2011) suggested that HR develop new capabilities and capacities to partner with finance and IT. We will show that in the human capital model in our case study, senior levels and other team members can easily capture and analyze the metrics of human capital management. HR team members share ownership and responsibility for these costs and the results, since they often set, enforce, and track the base data to produce the human capital costs. Too often HR has been silent on the costs of their actions, but by using the human capital metrics, they can see and feel the impact on the larger organization. HR always wanted a seat at the table and now can share in the financial impact of human resources policies.

Finally, it is important to address ethical concerns regarding HR and predictive analytics and ask ourselves when is it appropriate to apply HR analytics? This introduces ethical dilemmas that HR professionals might face. For example, is it ethical to identify employees or applicants and their prescription drug records or their probability of becoming violent at workplace? The literature review on this topic is limited. Falletta (2013) in a survey of 220 participants from Fortune 1000 companies found that 76% of the listed practices, in his research, were considered neutral or inappropriate by the respondents. In addition, it is quite likely that individual ethical judgments will vary depending on the variety of human capital decision from hiring to assessments, promotions, etc.

Big Data: Industry and Academics

The challenges of leveraging big data have compelled the industry and academics to work together and develop well-designed strategies to capitalize on the organization's data assets. Within the last 50 years, business schools have started to embrace sophisticated research methods, mathematical optimization, multivariate statistics, and econometrics to study marketing problems, such as customer behavior, lifetime value of a customer, etc., which later, combined with the emergence of data science and big data technologies, resulted in development of new professional organizations, scholarly journals, and specialized conferences and bringing teaching and practice together (Chintagunta et al. 2016). The hotel industry is not an exception. There have been a number of hotel chains that have utilized predictive analytics to make business decisions; however, these activities primarily address understanding customers and what they want from the hotel, customer analytics, loyalty marketing, and capacity and pricing optimization (http://duettocloud.com/resources/whitepapers/Bringing-Predictive-Analytics-to-the-Hotel-Industry.pdf). However, these analytics did not capture issues in human capital investment and returns.

Today, companies are targeting sufficient customer data that can provide valuable business information to support business decisions on par with financial and human resources (Sun, Cegielski, & Li, 2015). HR analytic companies have brought onboard a number of Ph.D. statisticians and business intelligence professionals and suggested involvement of academia in implementing analytical practices and case studies (Starner, 2017). Following aforementioned accomplishments in the marketing field, this case study opens up new doors and an example of a partnership between academics and industry to leverage the capabilities that each side can bring to the table as relate to HR. Figure 9.1 provides a brief narration of academic and industry partnership leading to this case study.

The Need for Human Resources Analytics in the Lodging Industry

For the HR field, to establish strategic and business partnerships, like the ones in the field of sales and marketing, it needs to become strategic partner with business. To achieve this goal, HR needs data-based decision-making and analytics capabilities to influence business strategy. This is crucial because if HR wants to play a strategic role in organizations, it needs to develop measures grounded on how human capital decisions affect the business and vice versa (Lawler, Levenson, & Boudreaau, 2004). For example, JetBlue, Best Buy, and Limited Brands have discovered an important statistical relationship between employee satisfaction and company performance (Davenport et al. 2010).

In the lodging industry like many other businesses, BI tools (platforms) can provide an architecture for businesses to analyze company data and reveal patterns and insights. These platforms offer a variety of solutions to HR challenges. A study conducted by G^2|crowd (2017) that recently rated BI tools (platforms) highlighted that "…to be qualified as BI platform a product must:

- Consume data from any source through file uploads, database querying, and application connectors
- Provide an architecture to transform data into a useful and relatable model

UD's Poorani, Sullivan introduce new metrics that apply human capital analytics to hospitality

University of Delaware hospitality professor Ali Poorani and Bill Sullivan, managing director of the Marriott Courtyard Newark at UD, have partnered to introduce new metrics for evaluating human capital performance in the lodging industry.

"The current metrics, such as average daily rates, occupancy percentages, revenue per available room, and various labor costs that hotels use to measure performance, though well-established, do not adequately quantify return on human capital investment," Sullivan said.

Nearly 90 percent of organizations report that leveraging large swaths of critical data, otherwise known as "big data," is critical to being competitive in their industries, Poorani added, but, "Hotels earn relatively low scores in utilizing big data and competitive intelligence. This is surprising."

Poorani said that he has been trying to apply the concept of human capital analytics to the lodging industry, but he had doubts that he would have access to the data, until he approached Sullivan.

Poorani and Sullivan then launched an investigation into the hotel industry. As a case study, they applied the Vienna Human Capital Performance Index, which uses various algorithms to calculate an organization's:

• Entire investment in human capital: Employee costs, costs in support of employees and costs in lieu of employees.

• Human capital productivity: The amount of revenue generated for each dollar invested in human capital after adjusting for the costs of material.

• Profit sensitivity: The ratio between profit driven incentives and profit goals determined by the organization.

The case study analyzed detailed financial data for the UD Marriott over a period of the past three years. (Article by Sunny Rosen)

http://www.delawarebusinesstimes.com/ud-hospitality-professor-creates-new-metrics-hotel-industry/

Fig. 9.1 Academic and industry partnership narrative

- Support data modeling, blending, and discovery processes
- Create reports and visualizations with business utility
- Create and deploy internal analytics applications"

For example, Sage Hospitality, a leading hotel company, began using ProfitSage Financial Suite as its business intelligence tool, since May 2008. The web-based ProfitSage software consists of six modules that interface with the PMS, back office accounting, point of sale, sales automation, and payroll systems in order to warehouse critical operational data. The information is then used within the different modules to produce revenue reports, detailed forecasts, line-by-line budgets and a number of other customized reports. ProfitSage also interfaces with Smith Travel reporting to provide

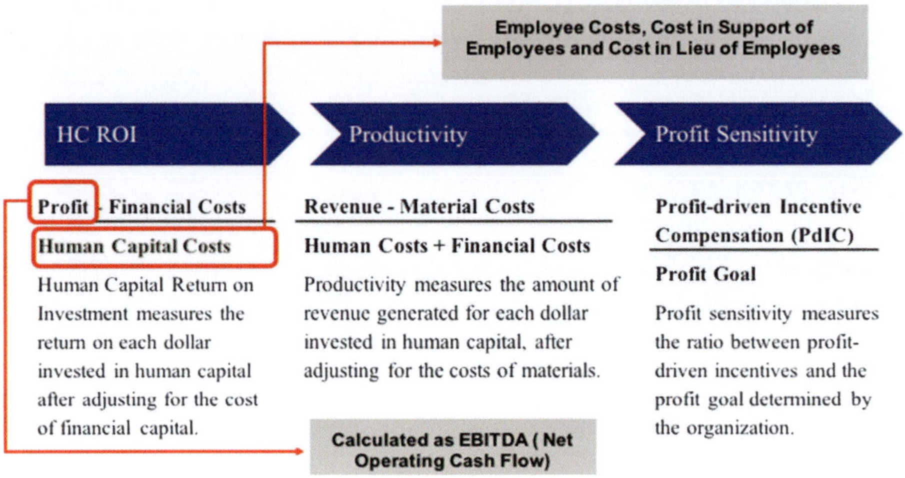

Fig. 9.2 Human capital return on investment, productivity, and profit sensitivity metrics

comprehensive comparisons to the competitive set (http://www.hotel-online.com/News/PR2008_3rd/Sept08_SageProfitSword.html). Similarly, Peachtree Hotel Group (PHG), one of the nation's fastest-growing hotel investment and management companies, has utilized ProfitSage applications for its portfolio of 34 hotels including Comfort, Starwood, Hyatt, Marriott, IHG, and Hilton brands (https://profitsword.com/peachtree-hotel-group-chooses-profitsword-for-business-intelligence-platform/).

In our case study, we used data driven from ProfitSage, a BI tool, to calculate the following metrics: return on investment, productivity, and efficiency of human capital. We hope as the process matures, the metrics discussed in this case study may become key performance indicators (KPIs) for lodging organization to better evaluate their investment in human capital.

HR Analytics BI Metrics

Figure 9.2 lists some of the most utilized metrics in HR adapted from Visier, Ultimate Software (white papers), DiBernardino and Miller (2008), and Ping (2013).

The Scope of HR Analytics

Generally, analytics for managing human capital can be categorized from simplest to most sophisticated in the following manner proposed by Davenport et al. (2010): *Human capital facts, analytical HR, human capital investment analysis, workforce forecasts, the talent value model, and the talent supply chain:*

1. **Human Capital Facts:** Explore a single truth regarding individual performance at the enterprise level, such as demographics, contingent labor use, turnover, or recruiting data. For example, Harrah's Entertainment uses insight from data to put the right employees in the right jobs.
2. **Analytical HR:** Analyzes individual performance data in relation to certain HR metrics. For example, comparing staff-turnover investigation in two distinct regions or how individual performance data relates to cost, time, engagement, or retention.
3. **Human Capital Investment:** Investigates which actions have the greatest impact on business outcome. For example, Sysco, a Fortune 100 global food-service company, has saved nearly $50 million in hiring and training costs by analyzing the company's delivery personnel and their satisfaction scores. When the satisfaction rates dip, the company institutes training and improvement to bring them back on track. As a result, the company has improved its retention rates from 65% to 85% in 6 years.
4. **Workforce Forecast:** Workforce planning involves an examination of a company's current and anticipated labor needs and predicting future gaps. It helps to avoid talent gaps by analyzing talent shortages, skill gaps, turnover, succession planning, and business opportunity data to identify potential shortages or excesses of key capabilities before they happen. For example, by analyzing historic data, one can determine why people are leaving, or does it cost more to retain employees or replace?
5. **The Talent Value Model:** Uses analytics to determine what the employees value the most or like about the company and choose to stay with the company. This can help the company to focus on certain individuals, groups, etc. to actively engage in their development. For example, Google uses this model by analyzing the employee performance data to help both high- and low-performing employees.
6. **The Talent Supply Chain:** Given the rapidly evolving workplace, it is essential that organizations maintain an effective talent supply chain. Big data helps, in real time, to determine the current talent, the talent the organization needs, or the talent that it will need in the near future. For example, a hotel can optimize next day's schedules based on forecasted sales patterns, inbound call center volumes, or predicted receipts. HR can help the organization to determine where a certain talent is needed to support the company strategy. In labor-incentive organizations, such as lodging industry, this can be critical for the long-term success of the organization.

What Is Needed in the Lodging Industry?

Lodging industry, also known as *hotel business*, encompasses various brands, types of hotels and motels, and related industries, such as bed and breakfasts, extended stay, Airbnb, and other providers. Marriott, with its recent merger with Starwood, is the leading provider of hotel services internationally.

All the six BI metrics listed above (the *scope of HR Analytics)* are essential for lodging industry, an industry that is heavily labor-incentive. However, in this case study, we will focus on *human capital investment metrics*, which investigate what actions and strategies have the greatest impact on business outcome (number 3 in the hierarchy proposed by Davenport et al. (2010).

To our knowledge, direct effects of human capital investment and its financial results have not yet been fully utilized in the lodging industry. To address this void, we used Vienna Human Capital Performance Index proposed by the Vienna Human Capital Advisors (DiBernardino & Miller, 2008) to investigate how the following three BI metrics, namely, *human capital ROI, productivity,* and *profit sensitivity* (see Fig. 9.2), can be utilized in the lodging industry to help management and HR to better justify any additional investment in human capital.

We propose that the three human capital metrics, *the human capital return on investment, productivity, and profit sensitivity*, signify innovative "human capital metrics," a set of new and possibly future key performance indicators (KPI's) for the lodging business. We believe that determining these measures are achievable, since a large volume of data are available for the lodging industry but not utilized in this manner. These new human capital metrics provide a new tool for management and HR to look at the targets that relate to the financial results of investment in human capital and a "going forward" way to separate human capital costs from other volumes of accounting data that are commonly used in the lodging industry, which will be discussed later in this chapter.

Key performance indicators are those metrics established by the firm as "critical success factors" for the business, and they are regularly updated and often highlighted in MIS or other dashboards for quick analysis of business results. The three metrics used in this case study will help analyze the root causes of issues relating to HR and measure success of investment in human capital based on targets set by the organization, or a competitive set. As a starting point, we will demonstrate how these models can become (KPIs) for the organization using a 3-year period of data analysis.

Value of Big Data in Lodging Industry

Biljihan and Wang (2016) stressed that in order to create a competitive advantage, lodging organizations need to integrate all new technologies and create synergy among human capital, knowledge, and financial capital. They argued that both academia and practitioners attempt to understand the driving factors of competitive intelligence. Academics and lodging businesses can benchmark KM, BI, and CI from other industries on a case-by-case basis and develop new predictive analysis to enhance the use of big data. Based on the authors' experience, companies often bring in consulting expertise to analyze operating data to better understand the figures and course of action based on that data analysis. Real data speaks for itself, but there is a need for further analysis by industry and educational experts to dig beyond the real data. The lodging industry needs to look at both short- and long-term trends and recommend actions to improve the results and even come up with new metrics to present data about human capital costs. The three metrics presented in this case study will show the business results of any additional dollars spent in human capital.

Methods

Case Studies

Case studies are powerful means of introducing the impact of analytics in the hospitality industry. For example, Vanneste (2016), by analyzing performance data in more than 400 of its restaurants in the United Kingdom, concluded: "McDonald's was able to ascertain that customer satisfaction averaged 20% higher in restaurants where employed staff are 60 years or older. An older market segment was drawn to the restaurants and the older employees had a positive impact on the team morale." In another case study, ConAgra Foods Inc. measured the impact of new leadership training by studying performance of 600 trained and 1600 untrained supervisors across 65 US-based plants to measure the effect of training on leader retention. The results indicated that the turnover rate of trained supervisors was around 5.3%, while the untrained supervisor turnover reached 11.4%, during the 12-month after training, resulting in $2.3 million saving in the first year of leadership initiative (Baron, 2016).

As relating to lodging industry a case study posted by Caruso (2013), Starwood Hotels (with more than 1000 hotels and 145,000 associates) used a predictive workforce analytics by gathering 360-degree feedback from 2700 of its leaders and established Starwood critical leadership competencies. Using viaPeople BI tool and its large database, Starwood concluded that leadership effectiveness leads to associate engagement, which leads to guest loyalty and consequently to higher RevPar Index (revenue per available room). This case study has critical implications to HR and underlines that certain leadership behaviors and competencies impact the company bottom-line and business results.

Issues, Controversies, and Problems

Lodging industry's human capital analytics often incorporates top-line financial metrics, focusing on evaluating past performances rather than incorporating the predictive power of data and their strategic impact. Among many financial ratios, hotel enterprises focus on the following key performance indicators, metrics, or ratios for measuring effectiveness and efficiency of their operations:

- Guest room occupancy = total guest rooms sold/total guest rooms available
- Average daily room rate = total rooms revenue/total guest rooms sold
- RevPar (revenue per available room) = total rooms revenue/total guest rooms available
- Average dining check = food revenue per meal period/total customers (covers served in that period)
- Operating efficiency ratio = gross operating profits/total revenue
- Return on investment = net income/total assets

The guest room occupancy, average daily room rates, and RevPar are data elements that are submitted by thousands of hotels on a daily basis to a neutral confidential third party, Smith Travel Research (https://www.strglobal.com), to permit peer comparisons (most commonly is known as comparative set analysis). The other data, such as total customers served, food revenue per meal, etc., are extracted from the hotel point of sale restaurant/bar system.

As it relates to human capital, the most used ratio is labor costs, calculated as a percentage of dollars spent on labor divided by total revenues. This ratio could be calculated for each department and/or for the whole hotel property. Obviously, these and other similar metrics provide some raw data for comparison; however they do not capture or predict overall contribution of human capital to organization's business outcomes. Other than labor hour and labor cost comparisons, nothing is close to the "human capital costs" as defined in the Vienna Human Performance Index (DiBernardino & Miller, 2008). In this case study, we use the Vienna Human Capital Performance Index to analyze *human capital return on investment (HC ROI), productivity, and profit sensitivity* for the proposed property, Newark, Courtyard Marriott.

Application

1. **Human capital return on investment (HC ROI)** uses the following metrics to calculate HC ROI:

$$\mathbf{HC\,ROI} = \frac{\mathbf{(Profit - financial\ capital\ costs)}}{\mathbf{Human\ capital\ costs}}$$

Human capital cost, defined as the metric to "Measure the organization's entire investment in human capital – *employee costs, costs in support of employees, and costs in lieu of employees.*"

Profit numbers are depicted from the income statement lines and earnings before interest, taxes, depreciation, and amortization (EBITDA) otherwise known as operating cash flows; financial capital costs include interest payments and cost of borrowing; and human capital costs include *employee costs, costs in support of employees, and costs in lieu of employees* as follows:

- Labor cost (wage and salary)
- Employee benefits costs (medical, 401 K, social security, education, relocation, sick pay, paid holidays, and employer contributions to healthcare benefits)
- Training and development, tuition reimbursement if applicable
- Recruiting costs (people, ads, development of position descriptions, etc.)
- Hiring costs (physicals, drug tests, background checks, reference checks, agency fees, referral fees from agencies)
- Orientation costs for new employees
- Uniforms issued by firm
- Employee meals and locker room costs

2. **Productivity** measures the amount of revenue generated for each dollar invested in human capital, after adjusting for the costs of materials, calculated as follows:

$$\textbf{Productivity} = \frac{(\textbf{Revenue} - \textbf{materials costs})}{(\textbf{Human capital costs} + \textbf{financial capital costs})}$$

This formula is an adaptation of the traditional financial measure for productivity (revenue ÷ assets), and it can normalize all kinds of enterprises by controlling for material costs (which vary greatly by industry). Most hotel and restaurant operations, given the service nature of the business, are labor intensive; and material costs are tightly monitored as a percentage of sales or by profit margin (sales minus direct costs of goods sold). The "productivity" metric takes the two asset categories (people and investment costs) and looks at the multiplier for revenue, after cost of goods sold (material cost). In this formula, the assumption is that material costs do not drive revenues, they are passed through costs, and only human capital and financial capital are the drivers of true revenues. This pinpoints the hotel's ability to seize productive assets and make a return.

3. **Profit Sensitivity** measures the ratio between profit driven incentive compensation (PdIC) and a profit goal determined by the organization:

$$\textbf{Profit sensitivity} = \frac{\textbf{Profit} - \textbf{driven incentive compensation}(\textbf{PdiC})}{\textbf{Profit goal}}$$

This formula is an adaptation of the quick ratio, also known as the acid test, used to measure liquidity. The quick ratio is the most stringent, current method used by finance professionals to ensure that liquidity levels are adequate enough to protect an organization's cash position. The Vienna Index's Profit Sensitivity indicator is a corollary of the acid test but with a laser focus on the organization's compensation structure. Managers have clear goals to gain incentive compensation (bonus), and this measures the percent that the reward is. This is a perfect addition to portfolio of metrics for any hotel leadership team.

Data Analysis

Using DiBernardino and Miller (2008) model of human capital analytics, we provide data for a 3-year period, 2014, 2015, and 2016, to learn how managers can readily measure return on human capital investment, productivity, and profit sensitivity and report the finding at the executive and property levels.

Human Capital Return on Investment (HC ROI)

We drilled down to financial data for the periods of 2013–2014, 2014–2015, and 2015–2016 and inserted the relevant data in formulas using an excel sheet. Our results are illustrated in Fig. 9.3. The findings indicate that the property's HC ROI has been negligibly negative (−) for the years 2013–2015 and hovering near (0.00) indicating that human capital had not added enough incremental value sufficient to

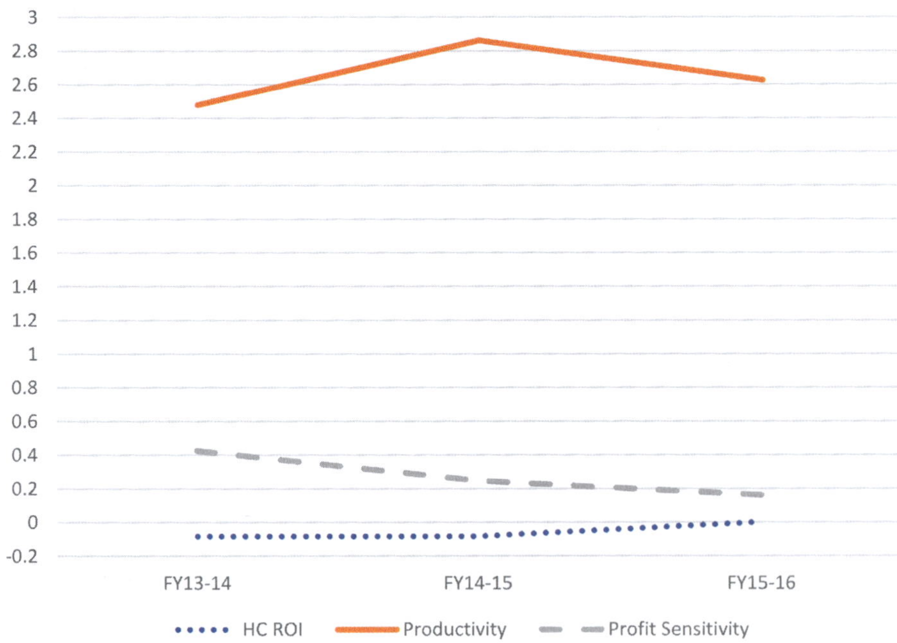

Fig. 9.3 Three-year trends for HC ROI, productivity, and profit sensitivity

cover financial capital costs. However, the trends show that the property is moving to a positive territory, during 2015–2016.

The 2015–2016 HC ROI shows that, in the most recent year, the property has achieved a positive territory and can allocate further investment in HC and still recoup the financial capital costs. The increase from FY14-15 to FY15-16, amounts to over 8% [(0.0029)–(−0.0831)], underlining that, keeping everything else constant, the hotel has gained $0.08 additional profit for each dollar invested in additional HC, justifying and, in some cases, defending future investment in HC.

Productivity

As mentioned earlier, productivity measures the amount of revenue generated for each dollar invested in human capital, after adjusting for cost of materials. In this formula, the assumption is that material costs do not drive revenues, they are passed through costs, and only human capital and financial capital are the drivers of true revenues. The results indicate positive and relatively stable productivity for the property hovering near 2.9X.

First, keeping other elements constant, an increase in revenue levels can boost productivity. Net Promoter Scores (Markey & Reichheld, 2009), guest surveys, number of repeat stays, retention (Joa, 2015), loyalty programs (Garnefeld, Eggert, Helm, & Tax, 2013), and customer sentiment analysis on social media (Leung, Law, Hoof, & Buhalis, 2013) are all reported to increase revenues. Courtyard Marriott, like most other hotel brands, uses all of the above. Second, many hotels routinely measure material costs. Management by controlling such costs can maintain a desirable ratio without sacrificing quality. Finally, a healthy productivity ratio provides flexibility to further invest in HC. Exploiting and propagating productivity ratios in the industry provide HR and management team to see the trends, compare the results over time, and benchmark with other similar establishments.

Profit Sensitivity

Profit sensitivity analytics helps organizations to establish a clear and measurable relationship between profit-driven incentive compensation (PdIC) and profit goals. A look at the 3-year history shows the volatility of profit-driven incentive compensation.

Based on our findings, as profit goals for this property increase, the percentage profit sensitivity decreases. By analyzing data over time and/or comparing this ratio to similar hotels, one can easily establish strategic target rates that are satisfactory for the organization and management, highlighting what percentage of profits could be dedicated to profit-driven incentives.

Studies have shown that profit sharing improves productivity, job satisfaction, organizational identification, and working hard (Bryson & Freeman, 2016; Fibirova & Petera, 2013). The metrics discussed here can help management to rightsize the profit goals and profit-driven incentives while maintaining liquidity to cover cost of capital.

Trend Analysis

As we see in Fig. 9.3, these measures provide a trend when using multiple years, which enables HR to clearly plan and predict what to expect in the future years. In the absence of available benchmarks in the lodging industry, the above measures are best utilized when the results are compared with the on-site established targets.

Obviously, the goal of exceeding the financial costs was not met in the year 2014 and 2015. This is not a surprise for lodging industry, in the early years, when the cost of debt is high. The data clarifies this expectation for strategic decisions. Also, it is not unusual in lodging enterprises to show smaller earnings as a percentage of sales due to the low margin of many hospitality businesses, especially where food and beverage sales are present.

On the upside, the trends show that our establishment has moved to the positive territory during 2016. HR and management can now analyze the historic data and evaluate what has contributed to this upward movement. As a result of data analytics, now HR can conclude that any additional dollar invested in HC can produce at the minimum, \$0.08 in profits, and the raw data show the dollar amount of additional investment available for HC, in the meantime, sufficient to cover the cost of capital.

The **productivity** metric is showing the trends for earnings turnover of the two cash assets (HC costs and capital costs) are increasing. This is clearly a step to more asset productivity and thus better earnings. Since capital costs are not controllable, fine-tuning the human capital costs for the better return is a win for all. In addition, reallocating human capital resources to address the areas deemed vital to guest service or adding resources from areas beyond human capital are also notable considerations.

Particular to the hotel and restaurant industry, material costs are significant. During hyperinflationary periods, the cost of material increases, and adjusting sales prices, accordingly, may be difficult in the short term. Productivity analysis can help to determine the root causes of diminishing productivity ratio. Accurate usage data, inventory control, preventing spoilage, etc. can also help to maintain productivity levels.

Profit sensitivity analysis can help to set clear targets for employee and management incentives. Obviously, driving higher returns while sharing the payoff with staff is one of the best motivation tools. Studies have argued profit sharing can result in improved productivity, job satisfaction, and value creation (Bryson & Freeman, 2016).

Reviewing the trends in the current case study, clearly the hotel trends are in the proper direction. The analysis shows the profits have substantially increased making the ration of incentives over profits smaller (the 2016 contribution to profit sharing hovers around 16% versus 40% in 2014). In absence of a stock sharing plan, the cash payout model is a real plus for the short-term drive needed to meet each year's business plan objectives.

Conclusion

The employee-centric focus is exemplified in the people-first culture at Marriott International. The enterprise's "Core Values and Heritage" web page clearly identifies valuing employees and providing associates with opportunities to grow and succeed is a top strategic priority. Quoting Executive Chairman, Bill Marriott, "Take care of associates and they will take care of the customers" (Marriott International, 2015).

The data, in this single property, provided strategic level understanding of this property's investment in human capital and its productivity levels for a 3-year period. Charting the three metrics of human capital ROI, productivity, and profit sensitivity over time for one hotel helped to show trends (positive, negative, and volatility) and establish baselines for the use of such metrics in the future.

Per Managing Director of the proposed property, "The exercise of this case study was an eye-opening experience, a change in mindset." Traditionally, accountants would slice and dice performance data extracted from financial data. This case study made it abundantly clear that management and HR can incorporate key performance and profitability data to improve organizational performance. The process further established that the BI tools used in this case study were relatively easy to use. The information was readily available and the management did not have to go through IT to perform such analytics. The analysis process was fairly simple using Excel sheet. The partnership with the University of Delaware faculty made the analysis more effective.

The process of manual calculations can even be further streamlined by building dashboards using industry-specific metrics to address key business outcomes and people strategies. The case study we developed and the BI methods we used will start a movement toward more sophisticated analytics that matter for bottom-line and business predictability.

Mindful about our tangible, positive experience and results from this case study, we plan to investigate additional case studies nationwide, in order to provide a platform that companies can benchmark their HC ROI, productivity, and profit sensitivity with similar properties and the industry averages. We also suggest the academia and lodging institutions to incorporate the value of other big data, such as Net Promoter Scores, guest surveys, number of repeat stays, memberships, loyalty program, etc., into financial metrics proposed in this case study. Clearly, feedback from other data sources can help human resources teams make changes in the recruiting process (such as redefining position description requirements or changing areas of recruitment), new employee and existing employee training, and other issues exposed in the process of financial returns in HC investment.

Appendix: Case Study Questions

Q1. How can HC analytics help us find where the most investments in HC should be?

Suggested answer: In calculating HC costs, we used all of the following figures:

- Labor cost (wage and salary)
- Employee benefits costs (medical, 401 K, social security, education, relocation, sick pay, paid holidays, and employer contributions to healthcare benefits)
- Training and development, tuition reimbursement if applicable
- Recruiting costs (people, ads, development of position descriptions, etc.)
- Hiring costs (physicals, drug tests, background checks, reference checks, agency fees, referral fees from agencies
- Orientation costs for new employees
- Uniforms issued by firm
- Employee meals and locker room costs

Management can scrutinize any of the above numbers and decide where the savings or additional investment is most appropriate. Obviously, in the lodging business, trade-offs between qualities of service, mainly customer satisfaction data and costs, could be considered first. As we look at the root causes of service issues, we can track back to various components of the human capital costs listed above. For example, costs in support of employees. We know that lack of familiarity with job procedures could be related to investment in training or turnover, both manageable costs. Likewise, corrective action could be taken to increase emphasis or investment in those areas. Each issue can be fully investigated to determine the best way to improve that measure. Management can decide the level of resource allocation based on the HC ROI results and connect it to other data sources considering the severity of the issues and overall impact on key guests' measures, such as "likelihood to return" or "likelihood to recommend."

Q2. Can tying sentiment analysis and guest satisfaction data to investments in people (human capital metrics) give us advance warnings of service delivery problems?

Suggested answer: In addition to direct guest feedback, most hotels have added feedback from user generated sites (Trip Advisor, Google, Yelp) to provide other sources of guest feedback. Marriott currently uses both own customer data and a social media index to analyze hotel performance against peer hotels. HC ROI metrics can expose areas of service improvement opportunity and enable management to focus on those areas. The related costs of that emphasis may well be visible in the component costs of human capital (i.e., training, pay, benefits, turnover). Mining the verbatim comments from these guest feedback sources is vital to getting to root cause, trends, and identification of possible corrective actions.

Q3. What kind of data can we bring into our data analytics tools and algorithms to better understand HC productivity?

Suggested answer: The productivity algorithm in this case study showed that, keeping other elements constant, an increase in revenue levels and reducing material costs can boost productivity ratio. To improve revenue levels companies can utilize Net Promoter Scores, guest surveys, number of repeat stays, loyalty programs, and customer sentiment analysis on social media to increase revenues. Similarly, companies by monitoring prevailing material costs and its trends can assure the target productivity levels and benchmark with other competitive set.

Q4. If we have service quality disruptions, how can we most effectively recover?

Suggested answer: Hotels need to prepare for more disruptions in the near future. Disruptions can come from both external and internal (on-site). From an external perspective, general state of economy, consumer mindset, technology, and business platform are critical (Deloitte: https://www2.deloitte.com/content/dam/Deloitte/us/Documents/consumer-business/us-cb-2017-travel-hospitality-industry-outlook.pdf). First, a strong economy can add to demand, which requires accurate workforce planning and recruitment effectiveness that can contribute to productivity. Reverse is the same when economy is not performing or some unexpected risks might be on horizon, such as spread of Zeka, etc. Second, consumer mindset and expectations are in state of flux. Companies need to get customized and personalized data in the hands of those who deliver service at all touch points in order to differentiate their business and service offerings to their clients using their customer behavior measures and social networks. Third, hotels are, to a degree, behind in delivering enabling technologies to their customers. For example, by utilizing innovative apps hotels can enable their customers to book rooms, make various appointments, control their room temperature, and communicate with staff or other guests. The Internet of Things (IoT), by connecting devices, offers numerous opportunities to hotel businesses, such as connecting customer smartphones to various sensors at the hotel to greet a guest by name upon arrival, lead them to their room, open the door automatically, and manage the other needs of customers based on their unique profile. Fourth, new business platforms will be needed as the industry's organic and vertical growth reaches maturation. That is, hotels engaging in mergers and acquisitions will need to add across travel experiences, such as retail, restaurants, local excursions, etc. to provide personalized experiences. Finally, HC metrics offered in this case study will be helpful to analyze, add people to service areas that have impact on revenue growth, and reduce them where they don't.

Q5. How can we link analytics to talent acquisition strategies? Retention? Engagement?

Suggested answer: Although this study focused on strategic level financial metrics of human capital, lodging industry faces many human capital challenges.

Other case studies and human capital analytics can help in a number of areas, such as new hires, employee post-hire performance, total reward systems, overtime analysis, and effective benefit plans. Additionally, turnover, resignation, and employee movement are common place in hotel industry. BI methods used in predictive human capital analytics can analyze available data across location, department, gender, age, promotion wait time, pay increase, training, and other factors and relate them to business outcomes.

Q6. What are potential ethical concerns with respect to leveraging analytics to measure human occipital performance?

Suggested answer: Management and HR need to be mindful of ethical considerations in using BI. Be wary of broad classifications of guests with trends in service issues. We cannot allow our staff to have preestablished perceptions of guests that will impact service and generate reactive behavior from guests (i.e., sensitivity to ADA conditions, race, gender, nationality). The privacy of guest feedback in all areas needs to be respected even if the guest has not taken reasonable precautions. HR predictive analytics tools can build employee profiles, such as engagement, dietary habits, shopping practices, tracking employee sentiments, etc. Companies are using employee data, such as who knows what? Who knows whom? Which employee is likely to quit? Data about surprise events or escalated incidents, etc. (Leong, 2017). Privacy of employees is a critical ethical consideration. Although these technologies are beneficial, companies need to create information governance committee (IG) to analyze their privacy culture, set guidelines, and encourage best practices.

Acknowledgment We would like to express our gratitude to the Vienna Human Capital Advisors, particularly Frank DiBernardino and Adrianne Mill who provided an innovative, clear, and well-defined description for the metrics used in this case study and for allowing us to apply these metrics to a real business case.

References

Aral, S., Brynjolfsson, E., & Wu, L. (2012). Three-way complementarities: Performance pay, human resource analytics, and information technology. *Management Science, 58*, 913–931.

Baron, R. (2016). *Human capital analytics: Understanding and optimizing the people impact of your organizations*. Retrieved from: https://www.slideshare.net/RickBaronSPHRSHRMSCP/human-capital-analytics-22016

Bassi, L. (2011). Raging debates in HR analytics. *People & Strategy, 34*(2), 14–18.

Biljihan, A., & Wang, Y. (2016). Technology induced competitive advantage: A case of lodging industry. *Journal of Hospitality and Tourism Technology, 7*(1), 37–59.

Bradlow, E., Gangwar, M., Kopalle, P., & Voleti, S. (2017). The role of big data and predictive analytics in retailing. *Journal of Retailing, 93*, 79–95.

Brands, K. (2014). Big data and business intelligence for management accountants. Strategic Finance. Retrieved from: http://go.galegroup.com/ps/i.do?p=AONE&sw=w&u=udel_main&v=2.1&id=GALE%7CA372250794&it=r&asid=4299e0335231e594a1ed2c0193c77e7f

Bryson, A., & Freeman, R. (2016). Profit sharing boosts employee productivity and satisfaction. *Harvard Business Review.* Retrieved from: https://hbr.org/2016/12/profit-sharing-boosts-employee-productivity-and-satisfaction

Caruso, K. (2013). *Case study: Starwood hotels uses predictive workforce analytics to link leadership behavior with business drivers*. Retrieved from http://web.viapeople.com/viaPeople-blog/bid/95118/Case-Study-Starwood-Hotels-Uses-Predictive-Workforce-Analytics-to-Link-Leadership-Behavior-with-Business-Drivers

Chintagunta, P., Hanssens, D., & Hauser, J. (2016). Marketing and data science: Together the future is ours. *Marketing Intelligence Review, 8*(2), 18–23.

Davenport, T., Harris, J., & Shapiro, J. (2010). Competing on talent analytics. *Harvard Business Review* Retrieved from: https://hbr.org/2010/10/competing-on-talent-analytics

DiBernardino, F., & Miller, A. (2008). Human capital analytics the missing link: Measuring financial returns on the human capital investment. Vienna human capital advisors. Retrieved from http://www.viennaindex.com/wp-content/uploads/2013/07/ExecutiveBriefing.pdf

Erikson, S., & Rothberg, H. (2015, September). Strategic approaches to knowledge and related intangibles. In *European Conference on Knowledge Management (p. 256). Academic Conferences International Limited.* Retrieved from http://search.proquest.com.udel.idm.oclc.org/docview/1728409043/fulltextPDF/DBBA87807ECA4295PQ/8?accountid=10457

Falletta, S. (2013). In search of HR intelligence: Evidence-based HR Analytics practices in high performing companies. *People & Strategy, 36*, 28–37.

Fibirova, J., & Petera, P. (2013). Profit-sharing – A tool for improving productivity, profitability and competitiveness of firms. *Journal of Competitiveness, 5*(4), 3–25.

G²|crowd. (2017). Business Intelligence (BI) platforms. Retrieved from https://www.g2crowd.com/grid_report/documents/grid-for-business-intelligence-platforms-winter-2017?gated_consumer=2ae9503e-6c27-4485-9284-38cb17dfe704&utm_campaign=gate-745430&tab=tab-id-314514--&tab-nested=tab-314364-2446-

Garnefeld, I., Eggert, A., Helm, S., & Tax, S. (2013). Growing existing Customers' revenue streams through customer referral programs. *Journal of Marketing, 77*(4), 17–32.

Halper, F. (2015). Next-generation analytics and platforms for business success. TDWA *Research.* Retrieved from http://www.pentaho.com/sites/default/files/uploads/resources/wp-tdwi-next-gen-analytics-platforms.pdf

Harris, J., Craig, E., & Light, D. (2011). Talent and analytics: New approaches, higher ROI. *Journal of Business Strategy, 32*, 4 13, 1.

Joa, J. (2015). Customer retention is king: The future of retention marketing. *Forbs.* Retrieved from https://www.forbes.com/sites/jerryjao/2015/01/21/customer-retention-is-king-retention-marketing-provides-greater-roi/#716abb6632cf

King, K. (2016). Data analytics in human resources: A case study and critical review. Human resource Development Review. Retrieved from http://journals.sagepub.com/doi/abs/10.1177/1534484316675818

Kline, S., & Harris, K. (2008). ROI is MIA: Why are hoteliers failing to demand the ROI of training? *International Journal of Contemporary Hospitality Management, 20*(1), 45–59.

Lawler III, E., & Boudreaau, J. (2015). *Global trends in human resources management: A twenty-year analysis*. Stanford, CA: Stanford University Press.

Lawler III, E., Levenson, A., & Boudreaau, J. (2004). HR metrics and analytics: Use and impact. Human resource planning. Retrieved from https://mgtinsights.files.wordpress.com/2010/10/hr-metrics-and-analytics-use-and-impact.pdf

Leong, K. (2017). Is your company using employee data ethically? *Harvard Business review.* Retrieved from https://hbr.org/2017/03/is-your-company-using-employee-data-ethically

Leung, D., Law, R., Hoof, H., & Buhalis, D. (2013). Social Media in Tourism and Hospitality: A literature review. *Journal of Travel & Tourism Marketing, 30*(1–2), 3–22.

Mandelbaum, R. (2014) *Examining hotel labor costs*. Retrieved from http://lodgingmagazine.com/examining-hotel-labor-costs/

Markey, R., & Reichheld, F. (2009). Closing the customer feedback loop. *Harvard Business Review, 87*(12), 43–47.

Marler, J., & Boudreau, J. (2017). An evidence-based review of HR analytics. *The International Journal of Human Resources Management, 28*(1), 3–26.

Marriott International Corporate Values. Retrieved on December 18th 2015., from www.marriott.com

Mayer-Schonberger, V., & Cukier, K. (2013). Big Data: A revolution that will transform how we live, work, and think. Research-Technology Management, Sept.-Oct. 2015, p. 66+. Retrieved from Academic OneFile, go.galegroup.com/ps/i.do?

McMillan, H. (2016). Addressing the 12 major challenges today's organizations face. *Journal of Applied management, 21*(2), 125–127.

Ping, E. (2013, September 7). *Strategic business intelligence for hr*. Retrieved from https://www.slideshare.net/PingElizabeth/strategic-business-intelligence-for-hr-6-hr-metrics-no-executive-should-be-without-by-ultimate-software

Starner. (2017). HR analytics trailblazers. H*uman Resources Executive*, 14–16. Retrieved from http://www.hreonline.com/HRE/view/story.jhtml?id=534362176

Stevens, L. (2015, March 9). Taking analytics up a notch. *Human Resource Executive Online.*

Sun, S., Cegielski, C., & Li, Z. (2015). Research note for amassing and analyzing customer data in the age of big data: A case study of Haier's online-to-offline (O2O) business model. *Journal of Information Technology, 17*(3/4), 166–171.

Tracey, J. B., & Hinkin, T. (2006, December 01) *The costs of employee turnover: When the devil is in the details*. Retrieved from http://scholarship.sha.cornell.edu/cgi/viewcontent.cgi?article=1148&context=chrpubs

Vanneste. (2016). Retrieved from https://home.kpmg.com/be/en/home/insights/2016/11/the-datafication-of-hr-bernd-carette.html

Visier. (n.d.). (Whitepaper) Retrieved from: https://www.visier.com/lp/wp-datafication-of-hr/

Part III
Aligning Strategies and Business Analytics

Chapter 10
Delivering Internal Business Intelligence Services: How Different Strategies Allow Companies to Succeed by Failing Fast

Rubén A. Mendoza

Abstract This chapter reviews opportunities and issues propelling and limiting the success of business intelligence and analytics services for a company's internal use. We describe three strategies for providing these services internally (on-premises, cloud, and hybrid) and explore issues of importance in the shaping of current demand and of future offerings by web-based providers. It also discusses opportunities for the development of academic curricula to offer better training to graduate and improve recruiting outcomes for organizations and for the development of more relevant academic research to address topics of current and strategic importance to the firm.

Keywords Business intelligence · Business analytics · Cloud · Service architectures

Introduction

Business intelligence and analytics (BI&A) are a category of computing technologies and supporting processes facilitating data collection, storage, access, and analysis to improve managerial decision-making (Chaudhuri, Dayal, & Narasayya, 2011; Watson, 2009). *Business intelligence* generally refers to technologies and processes which provide *descriptive* understanding of data sets, from non-interactive reports to dynamic dashboards capable of drill-down activities. Front-end tools such as Tableau, Qlik, Cognos, and others provide multiple configuration options to create stunning, interactive data visualizations. At the back-end, high-performance databases with support for relational, object, and non-relational data capable of storing and accessing traditional structured data as well as unstructured multiformat data are critical to BI activities. BI product options abound, but Microsoft SQL Server, Oracle Enterprise Server, IBM DB2, SAP ASE, and MySQL are leading choices.

R. A. Mendoza (✉)
Saint Joseph's University, Philadelphia, PA, USA
e-mail: rmendoza@sju.edu

M. Anandarajan, T. D. Harrison (eds.), *Aligning Business Strategies and Analytics*, Advances in Analytics and Data Science 1,
https://doi.org/10.1007/978-3-319-93299-6_10

Business analytics (also commonly known as data analytics) is generally used to describe higher-level statistical analysis of large data sets. While data analysis is necessary to the development of a story conveyed via BI reports, the value of data analytics lies in its predictive and prescriptive nature. Predictive analytics uses sophisticated data modeling to make forecasts, and prescriptive analytics makes specific recommendations on an optimal course of action (Davenport, 2013). In addition, business analytics generally assumes self-service activities, in which the analysis is performed by business users who are also specialists in data analysis techniques. Business analytics is closer to process than to product, but many (if not all) of the products listed above support analytics processes adequately.

BI&A success stories have more to do with the company's innovative use of existing tools than with the capabilities of the tools themselves. The opportunities are endless: the Burberry Group personalizes in-store customer shopping experiences with the use of RFID tags and targeted videos on display screens; CVS Health matches call center agents with the types of customers agents are more likely to engage positively; L'Oreal mines social media posts and routes appropriate ones to employees who can engage with the post's author; Lockheed Martin mines project data for early indicators of trouble; and Petroleos Mexicanos uses sensors to monitor equipment noise to proactively schedule maintenance and reduce downtime (Laskowski, 2015). For simplicity, the term "BI" will be used here to refer to any and all technologies and processes described here as included in BI&A.

This chapter provides a discussion of the three most commonly adopted approaches to the provision of internal BI services by organizations: on-premises, on the cloud, and a hybrid approach of the two. A description of some common benefits and limitations of these three common strategies on the spectrum of BI service provisioning accompanies each description.

An on-premise strategy involves the use of the firm's own human and technology assets to complete all business analytics activities resulting in any BI product. At the complete opposite end of the spectrum, a pure cloud strategy requires the use of external, network-accessible vendors, commonly referred to as cloud providers for all BI activities. The description of this strategy will also be accompanied by a discussion of its defining characteristics, primary service models, and common private deployment modalities. Finally, in a hybrid strategy, companies offer BI services using a combination of its own existing infrastructure and cloud providers.

Business Intelligence and Analytics

What constitutes a product or service in the business intelligence and analytics space is still relatively open to debate and can include anything from static reports (printed or electronic) to mobile, customizable/configurable, and on-demand data visualizations with real-time data and automated insight discovery. The simplest BI&A tools provide a summary view of historical data, often without the ability to request additional context for the data. Farther along the spectrum of complexity,

Apoteket, a state-owned Swedish pharmacy retailer, uses dashboard applications to analyze marketing data to identify effective store promotions, track the performance of products via online orders, and coordinate supply chains with pharmaceutical manufacturers and other retail partners (Tableau.com). The technologies which make this possible are, generally, familiar technologies: large, high-performance databases, telecommunications networks, mobile devices, and user-friendly interfaces. The business processes they support are also very familiar: identify profitable customers, products, services, and market opportunities, reduce costs, and increase customer satisfaction. The combination of increasingly powerful hardware/software platforms and business insight is what drives the innovation we see in BI&A.

Business intelligence and analytics have been described as both process and product: the results of processes and activities firms use to identify and extract useful information become the product which allows them to compete more successfully (Jourdan, Rainer, & Marshall, 2008; Vedder, Vanecek, Guynes, & Cappel, 1999). The process itself becomes the product as it generates insight. As processes mature and become proven industry bench marks, they are embedded in new software products as standard capabilities. The process becomes replicable and, eventually, generic (Carr, 2003). BI&A technologies and processes only provide as much competitive advantage as the firm can extract from data. Maintaining lasting competitive advantage becomes difficult because the creation of in-house applications specific to the firm's needs is expensive and slow, and off-the-shelf products are one size fits all. Firms whose BI processes or technologies are unique (likelier with on-premise strategy) may find longer-lasting positions of competitive advantage (Barney, 1991). However, cloud platforms are ideal for fast innovation cycles because service elasticity dramatically lowers the risk of provisioning a service incorrectly (Armbrust et al., 2009). As such, a unified framework for identifying a *single* corporate strategy for provisioning internal BI services either in-house, on the cloud, or in a combination of both is unlikely to emerge soon.

While many academic and industry researchers do not bother to differentiate between business intelligence and analytics (Chen, Chiang, & Storey, 2012; Gartner, 2013), others describe analytics as extending beyond large-scale data storage, mining, and colorful interactive visualizations to develop new products and services based on deep analysis of all the data companies maintain (Davenport, 2006, 2013). The notion of business intelligence has been around since at least the late 1950s (Luhn, 1958), but it was not popular with the business and IT communities until the late 1990s (Chen et al. 2012). At that point in time, *business intelligence* included all processes and technologies in support of better decision-making. In the late 2000s, the added *& analytics* descriptor began to make its way into research and industry literature (Davenport, 2006). In the 2010s, the two terms have started defining separate but related ideas, and while there is still not a consensus, uniform definition of each, differences are starting to become clear.

Chen et al. (2012), in their introduction to a special issue of the MIS Quarterly, the premier MIS academic research outlet (Lowry et al., 2013), synthesize a number of academic and industry articles and further classify BI&A into generations given the ever popular 1.0, 2.0, and 3.0 labels. According to the authors, BI&A 1.0

includes most current technologies, procedures, and uses, which rely on the collection of structured data via transactional systems (often legacy) and store it in traditional relational database environments. Data analysis is based on techniques developed in the 1970s and 1980s, consisting largely of intuitive, simple visualizations. BI&A 1.0 examples include practically any static data report or visualization but can extend to more current practices with dashboard applications. From spreadsheets to most OLAP activities, BI&A 1.0 captures what most companies are currently doing with their data.

BI&A 2.0 technologies and activities add text mining, web analytics, and social network analysis to the mix of data analysis activities discussed in BI&A 1.0. Perhaps the most recognizable label associated with BI&A 2.0 is *Big data*. Big data has three defining characteristics: volume, velocity, and variety (McAfee & Brynjolfsson, 2012). The *volume* of data being created by today's business environment is truly spectacular. And it is being created at an astonishing rate of speed (*velocity*). It is one thing to generate one petabyte of data (one million gigabytes, or 10^{15} bytes), but it is quite another to generate it in a single day. It is estimated that in 2016 Netflix users watched around 8.3 petabytes of video per hour and that total consumption of data in the United States topped 2.5 petabytes *per minute* (James, 2016). The *variety* of big data is reflected in the multitude of sizes, formats, and contents of the data being created and includes both structured and unstructured data. Structured data is most easily described as the kind of predictable data associated with online purchases: customer name, shipping address, credit card information, etc. It is predictable because all data fields are easy to categorize, have a range of anticipated values and well-known formats, and are subject to maximum sizes which do not sacrifice content or meaning. Building a database to accommodate this data is a simpler task (as difficult as building a good database is). Unstructured data is normally defined to include multimedia, graphics, e-mail, and a large number of social media content formats (Tallon, Short, & Harkins, 2013) which is much less suited to simple numerical analysis or categorization (Devlin, 2010). To illustrate that last point, the reader is invited to consider what the average of his/her last 20 social media posts would be (the question itself makes little sense and invites more questions in order to define it better). Building data repositories for unstructured data is a more complex task because it is much more difficult to predict its content and to categorize it. Extracting meaning from data with such varied origins, content, and format is a formidable challenge.

Lastly, BI&A 3.0 is the label applied to data products with analytics embedded, which provide and create data simultaneously. These technologies and activities support and enable location-aware, mass-customized, context-sensitive, mobile analysis and product/service delivery. For example, location-relevant advertisements and discounts in mobile traffic apps such as Waze, which can alert you to the presence of establishments or offers in your vicinity (data delivery) and record and report whether you take advantage of their offers (data creation). Using Yelp, consumers can check into restaurants and receive special discounts on the spot (data delivery and creation) and request a review of their experience later on (data creation). Other offers which consume and create data are tasks in apps like LinkedIn via in-network endorsements or recommendations, or the ability to modify Facebook ad campaigns in real time (as an advertiser) and to request that similar ads not be shown to you (as consumer) are all BI&A 3.0 services. Davenport (2013) calls them simply Analytics 3.0.

It is also worth noting that there is overlap and bleed at the edges of BI&A 1.0, 2.0, and 3.0 services, and a clear-cut border is often difficult to find. One is also likely to see a blurring of the lines between the terms business intelligence and analytics in current environments. In general, however, an on-prem approach is adequate for BI 1.0 activities, a hybrid approach can complement on-prem strategies to provide BI 2.0 services quite well, and the speed of application development in a cloud strategy is better suited for the intensive platform demands of BI 3.0 applications and products.

The Society for Information Management's (SIM) long-running (since the 1980s) IT *Issues and Trends* study series shows BI has remained at or near the top of technology investment by surveyed organizations in the United States for over a decade (Luftman, Kempaiah, & Nash, 2005, many others) and in recent surveys across several other geographical regions as well (Luftman et al., 2013). Similarly, IT and business alignment has been on the list of top five management concerns for an even longer period of time (Kappelman et al., 2015; Luftman & McLean, 2004; many others). In the past few years, the SIM survey has measured cloud computing and its many associated manifestations (Xaas, SOA, etc.) separately, and they have been shown to be among the areas receiving the highest levels of investment by organizations (Kappelman et al., 2015; Luftman & Derksen, 2012). Combined, all these results point toward the importance of aligning BI service capabilities with business objectives, especially when cloud computing activities are receiving such high level of attention and investment.

BI Service Architecture

A discussion of a general architecture to support the delivery of internal BI services through various modalities follows. By necessity, BI architectures will vary based on business need, budget, and technology capability (Shariat & Hightower, 2007; Turban, Sharda, Aronson, & King, 2008; Watson, 2009; many others) and will contain differing combinations of data sources, storage, and reporting technologies (Ong, Siew, & Wong, 2011). Figure 10.1 illustrates a generic BI services architecture suitable for the discussion of provisioning of internal BI services in this chapter.

Starting from the left side of the diagram, Fig. 10.1 shows the two possible sources of data (*internal* and *external*) and some of the various combinations of systems and partners from which data can originate. Data may be generated internally by transactional systems which collect data as customers order products or services from companies (TPS), as products move through a supply chain (SCM, ERP), as customer data is collected by sales staff (CRM), as raw materials are received and processed, and as products are manufactured, shipped out, etc. (op sys).

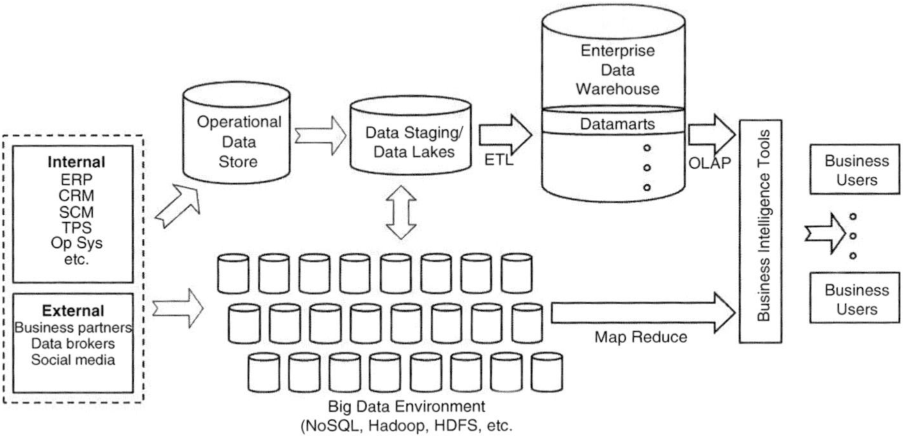

Fig. 10.1 Business intelligence architecture

Data may also be generated externally as a firm purchases marketing data sets, tracks product movement through the supply chain, advertises on social media outlets, monitors its online presence, engages with customers via social media, etc. As the data is generated internally or received from external sources, the firm may store it temporarily in an *operational data store* (not much more than a temporary database) until any preliminary data formatting, validation, and other activities are performed or until a suitable time after which the data may be moved elsewhere for preparation. This could be the end of a transaction, close of business day, the end of an online campaign, or once a predetermined amount of data has been collected. The operational data store shown is a proxy for various possible collection points. Though Fig. 10.1 does not explicitly indicate it, this data may be analyzed for near real-time trends and insights, but companies must be careful about interpreting results at this stage of analysis because data may be incomplete, redundant, or inaccurate and produce flawed insights (van der Lans, 2009a, 2009b, 2009c).

In preparation for eventual storage in an enterprise data warehouse, data may be moved to a *data staging* location (to make space for new data coming into the operational data store). With the rise of big data, this data staging location may also be known as a *data lake* and be used to combine structured data typically stored in relational environments with unstructured data coming from external environments. These environments will most frequently be implemented using well-known big data frameworks, technologies, and tools such as NoSQL databases, Hadoop, HDFS, and many others. The data in these staging areas or lakes can undergo analysis or additional and extensive filtering, validation, reformatting, and selection, collectively known as *extract-transform-load* (*ETL*; Watson, 2009).

An *enterprise data warehouse* (*EDW*) is a long-term storage and access environment for the data. Transactional systems generate and collect the data to be stored in the EDW and are dynamic and volatile, i.e., data changes very quickly. An EDW, by contrast, is static and batch-oriented, and ETL activities are critical to bridging the gap between the low complexity of the data processing in operational data stores and the very high complexity and accuracy required of long-term EDW storage to support BI services (DeLua, 2008). EDW contents are often categorized into data marts, collections of data specifically suited for targeted use (Chen et al. 2012). For example, some of the data contents of an EDW may be flagged as being of particular interest to the marketing or accounting functions, and some of this data may be of interest to both. This can be done with or without the need for data replication, and any given piece of data may be part of multiple data marts.

Online analytical processing (OLAP; Chen et al. 2012) tools and activities feed data directly to users or to additional BI tools such as dashboards, predictive modeling and data mining engines, etc. Big data mechanisms (such as MapReduce) can also provide users with data stored in distributed environments (like Hadoop) for direct analysis.

This simple architecture can be made more complex by the addition of technologies and processes designed to help the organization define data access and system configuration globally (master data management), specify, standardize, and codify the meaning of any data structure (metadata), facilitate the exchange of data between systems on demand (enterprise data and messaging buses), engage software services within and outside the company (web services, SOA), etc. The basic structure presented in Fig. 10.1 is sufficient to guide the discussion in this chapter.

BI Service Provisioning Strategies

This section discusses three strategies for the provisioning of internal BI services, beginning with the full ownership of all assets needed by the company itself, using its own technology infrastructure, followed by the opposite end of the spectrum, a fully cloud-based approach. A discussion of a hybrid strategy closes out the section.

On-Prem Strategy

The provisioning of a firm's internal BI services using its own human and technology assets is commonly known as *on-premises*, for obvious reasons. The term is hereafter shortened to *on-prem* for convenience and to reflect common industry practice. A pure on-prem strategy requires that a company own *all human and technology assets* required to create, collect, store, and analyze data, on both the client and on the server side, including all network assets in physical locations where the firm's employees are permanently located. In anticipation of the discussion of a hybrid provisioning strategy, it is important to note that we still define remote access to on-prem BI assets as on-prem, because the BI assets *themselves* are the firm's property. Among the benefits of a pure on-prem strategy are the development of expertise with the planning, implementation, operation, and replacement of traditional IT infrastructures, the ability to completely specify all security controls and environments for all BI services, and greater business agility/service speed. These benefits, and how they are accrued, are discussed below.

First, the ability to offer BI services on-prem allows an organization to develop in-house expertise with all aspects of information systems service delivery, from implementation to operation and eventual replacement, enhancing the firm's ability to recognize the value of technical innovations, an idea labeled absorptive capacity (Cohen & Levinthal, 1990). Zahra and George (2002) extended the concept to differentiate between potential and realized capacity and defined them as the acquisition and assimilation of new information or innovations and the transformation and exploitation of either or both. Additional research argues that competency with BI systems increases the firm's ability to acquire and transform additional innovations and to assimilate BI technologies and services (Yeoh, Richards, & Wang, 2013). Empirical work by ElBashir, Collier, and Sutton (2011) shows that, while top management support is a strong moderator of the deployment of BI technologies and services, the power to assimilate and transform BI capabilities comes "from the bottom-up" (p. 180), that is, from the managers working directly with these services and who are closer to the operations of the firm. On-prem delivery of BI services can be a tremendous advantage to the corporation.

Second, security and privacy and their associated controls have been top-level strategic concerns for management for more than 10 years (Luftman & McLean, 2004; Kappelman et al., 2015; many others). These concerns will remain at the forefront of thought for strategic planners as IT service provisioning priorities shift from tactical and operational concerns to strategic and value-generating areas (Kappelman et al., 2015). The ability to completely regulate the security controls environment for BI services is a very attractive benefit of on-prem provisioning. This is an increasingly more difficult assignment for any organization, and total control over data security and privacy is very much desirable.

Lastly, the annual SIM industry survey previously cited has also identified business agility as a top management concern for well over a decade. Business agility has been defined variously as the ability to react to changing market conditions more effectively and efficiently and to identify prospects for innovation, maintain superior customer satisfaction, and gather expertise, technology, partnerships, and other assets to act on those opportunities with speed and surprise (D'Aveni, 1994; Goldman, Nagel, & Preiss, 1995). Wixom, Yen, and Relich (2013) expand that definition by including the speed at which an organization can turn raw data into useful information, an ability they term "speed to insight" (Wixom et al., 2013). A firm's expertise with on-prem BI services can contribute to business agility by enabling truly customized service development with in-house expertise.

Cloud Strategy

The second BI strategy discussed here is the complete provisioning of all BI services via external, network-accessible vendors, commonly referred to as cloud providers. Fittingly, this approach is referred to here as, simply, cloud. Before a description of what cloud services are and could be, it is worth noting that cloud services are defined by both providers and customers as their needs require it, and no single framework theoretical to explain their adoption, diffusion, or assimilation yet exists. The reasons for the corporate adoption of any cloud service are as varied as the services themselves. Whether any particular company derives any benefit from any given cloud service is so heavily dependent on context, as to make a universal framework of analysis a daunting task. This section provides a brief history of the development of cloud services and descriptions of their defining characteristics, service offer models, and customer deployment models and closes with a brief discussion of some of their benefits and limits.

The provisioning of computing services over network connections by third-party providers has a checkered past. Companies looking for cost savings began to outsource computing and software services in the early 1990s, and as network bandwidth began to increase and costs decreased, access to those services over network connections became much more attractive options. The first generation of this type of service involved access to commercial off-the-shelf software on a subscription or on-demand basis, and vendors became known as application service providers (ASPs; Lee, Kim, & Kim, 2007). The initial success of ASPs was propelled by a desire for firms to concentrate on their core competencies, reduced IT costs, and a shortage of skilled workers (Lee et al. 2007). The collapse of stock market prices for Internet pure plays in the late 1990s led to the failure or consolidation of most "dot-coms" and to the restructuring of Internet-based business models. After that period of time, the ASP acronym fell out of favor due to its association with the dot-com bust. The outsourcing of software services model, however, survived and reemerged with several variants and names, such as Software as a Service (Saas) and the preferred label of cloud computing we use today, which has now grown to include SaaS. Cloud computing has been formally defined by the US Department of Commerce's National Institute of Standards and Technology (NIST) as:

> enabling ubiquitous, convenient, on-demand network access to a shared pool of configurable computing resources (e.g., networks, servers, storage, applications, and services) that can be rapidly provisioned and released with minimal management effort or service provider interaction. (Mell & Grance, 2011, p. 2)

This shared pool of resources has five defining characteristics:

1. On-demand self-service, such that consumers can engage services as needed, when needed, without human intervention
2. Broad network access over standard telecommunications technologies for heterogeneous computing platforms

3. Resource pooling, otherwise known as multi-tenancy, in which numerous service consumers share hardware, software, and other resources, dynamically assigned according to demand, and without the consumer having knowledge or control over the physical location of said resources
4. Rapid elasticity, which gives the impression of unlimited resources to consumers who can scale demand up or down without prior negotiation
5. Measured service, which helps optimize the allocation of resources by providers and charges consumers only for what they use

Cloud services are generally delivered in one of three primary service models: Software, Platform, and Infrastructure as a Service (SaaS, PaaS, and IaaS, respectively). The differences can be subtle, but it is largely accurate to say SaaS is software, PaaS is hardware, and IaaS is both. Thus, consumers have a range of options over which to engage software services offered by the provider, use the provider's hardware to run the consumer's choice of software, or use the provider's hardware/software platform as if owned by the consumer.

Deployment models range from completely private clouds to shared (community cloud), public, or combinations of the above (hybrid cloud). While a private cloud sounds suspiciously like on-prem service (and does, in fact, include it), the resources on the cloud may be owned and operated by a third party, on- or off premises. Thus, a careful distinction must be made (however fuzzy it may become) between how the company *chooses* to provide internal BI services and how the services are *offered*. In this chapter, any portion of an internal BI service not owned and operated by the consuming organization is considered a hybrid strategy, which could be offered in a private, community, public, or hybrid cloud environment.

In a pure cloud strategy, companies perform all BI activities, from collection to storage to analysis, using computing resources and analytics software offered by network-based providers (Thompson, 2009). The rise of big data and the increasing affordability of processing and storage have been major drivers for the rise of cloud BI. As the volume of data generated by firms and the affordability of computing power both grow, more and more companies are finding it attractive to outsource BI services to external vendors. Many large organizations not traditionally in either the technology or technology services sectors now offer access to their vast computing and storage resources to provide almost any level of service to numerous clients. Companies like Amazon, Google, IBM, and Microsoft have been consistently among the industry leaders in cloud services for several years (Darrow, 2015). These established firms can leverage existing technology investment and expertise in IT/IS service provisioning and generate more market opportunities for themselves. Newer players such as CenturyLink, 1&1, Rackspace, and a multitude of smaller national and regional cloud providers (Liu, 2016) compete vigorously in what is sometimes referred to as the service as a service (LeMerle, 2012) arena, which also uses the SaaS acronym. Others prefer the term everything as a service, abbreviated as XaaS.

Gartner's Magic Quadrant reports describe a dizzying array of services from a number of players, including professional services companies like PricewaterhouseCoopers, KPMG, Ernst & Young, Deloitte, traditional software developers like SAP, and international players such as Tata, Wipro, and Tech Mahindra (Heizenberg, Lo, & Chandler, 2017).

Some of the benefits of a cloud strategy are greater business agility, scalability, platform flexibility, enhanced security, and cost advantages. *Agility* is a firm's ability to respond effectively and quickly to unanticipated change (Goldman et al. 1995), and a cloud strategy enables companies to experiment with new BI services without undergoing infrastructure modifications and committing extensive resources (capital, technology, or human). In turn, this allows them to "fail fast" and learn about useful BI service features in shorter cycles. *Scalability* is the capacity to grow services on a large scale quickly and in consistent fashion and then be drawn down to meet lower demand without sacrificing performance or compromising capital investment. *Platform flexibility* is the ability to quickly take advantage of multiple platforms offered by service providers, which allows companies to generate greater numbers of digital service options (Sambamurthy, Bharadwaj, & Grover, 2003) to develop new products and services and increase speed to market. Business continuity and security controls offered by service providers are critical, and companies may choose tested and certified offerings which match or exceed their own ability to provide *enhanced security*. Lastly, *cost advantages* include the on-demand nature of these services and of their fee structures, which remove capital investment constraints for firms and enable targeted investment in BI services with quick ROI.

Among the drawbacks of a cloud strategy, one may consider the loss of opportunity to develop in-house technical expertise with the technologies required to provide BI services, a trade-off between complete customization and limited configuration based on provider capabilities, dependence on external security controls environments, and the potential for more difficult egress from a service provider due to closed data architectures and formats.

Hybrid Strategy

In a hybrid strategy, companies provide some BI services using existing on-prem infrastructure, and other BI services using cloud providers. There are almost limitless combinations of on-prem and cloud combinations to choose from, and they can start with very small expenditures. Firms may maintain complete control of data collection and only outsource analysis, or do any of the data collection, storage, access, or analysis on the cloud and complete the rest of the tasks on-prem. A company may even choose a different approach for its different business units, product lines, or physical locations.

Not surprisingly, a hybrid approach may include all benefits and drawbacks typically found in on-prem and cloud strategies, and not necessarily in best-of-both-worlds combinations. A company may find the agility provided via cloud services to be a tremendous opportunity but fail to develop a strategic base of expertise in-house by providing its BI services externally. At the same time, a company could be in complete control of its cloud-based data access and controls environment but choose a technology platform which makes it extremely difficult to switch providers later in time.

The next section illustrates successes in the delivery of internal BI services using on-prem, cloud, and hybrid approaches and issues limiting their growth.

Industry Opportunities and Issues

A review of current practices and industry reports shows a number of practices driving the success of internal BI service delivery by organizations of various sizes and scopes (Heizenberg et al. 2017). Many service providers deftly integrate business advisory expertise with deep and/or wide-ranging technical skills to successfully develop appropriate service platforms for their clients. IBM is noted for its wide BI product capabilities and its history of success with its clients.

Service providers which emphasize cloud-based over on-prem solutions have a wider array of options to offer clients, along with a range of attractive pricing models. Oracle is offered as an example of a company with wide-ranging on-prem and cloud solutions aimed at making the transition to cloud-based services easier. Amazon's web services platform (AWS.com) is one of the dominant players in the cloud arena, partly because of its many product choices, and the flexible pay-as-you-go options for its offerings.

The largest professional services organizations, also known as the Big Four (Deloitte, Ernst & Young, KPMG, PwC), offer services on various industries, supporting their central staff with international resources from a strong network of member firms. The Big Four professional services firms have extensive global networks: Deloitte employs more than 240,000 people in 150+ countries, EY (ey.com) reports over 230,000 professionals in over 150 countries, KPMG (kpmg.com) has over 174,000 people in 155 countries, and PwC (pwc.com) employs more than 220,000 associates in 157 countries. Partnerships with local resources can lead to innovation and service upsell due to the providers' ability to deliver locally on projects of greater complexity, involving multiple technical platforms, lines of business, and locations (Heizenberg et al. 2017). Large organizations with many capable partners can deliver sophisticated service portfolios focused on a number of different industries and become trusted strategic partners for organizations. There are opportunities for smaller providers who can be nimble and respond faster to specific needs. Companies such as KPI Partners, CBIG Consulting, Protiviti, and many others do not have the number of employees or the global network of partners the Big Four have but offer well-regarded services.

A number of common threads regarding barriers to the success of internal BI service provision emerge. The 2017 Gartner Magic Quadrant report shows several of their top-ranked business intelligence and analytics vendors possess excellent business advisory depth or deep technical skills. The report also states deep expertise in both areas is much less common. Other challenges range from issues with system integration expertise, to lack of depth "on the bench," leading to scarcity of knowledgeable technical experts to respond quickly enough to challenges. On the business advisory side, thought leadership, simple consulting skills, project and change management, or innovative solutions are listed by clients as commonly found barriers.

Service provider size can compound these problems. Too small, and providers are often unable to balance staff deployment with needs, provide quick turnaround, or satisfy demand for domain expertise. Gartner cites evidence that regional providers face problems with consistency of execution, methods, or best practices across geographies. For example, the 2017 Magic Quadrant report states regional European, Canadian, and Indian providers and system integrators frequently deal with inconsistent engagement scope and bench marks, lower-than-expected business deliverables, better regional than global results, and cultural fit issues.

Large providers can become riddled with bureaucratic procedures, coordination issues across locations or business units, and still be spread too thin, and regardless of size, the greater the complexity of the projects providers take on, the higher the price tag for organizations.

Vendors with deep technology expertise face issues beyond project or change management challenges. For organizations with deep technology expertise with one or a few platforms, or whose BI services and solutions are based on their own proprietary platforms and applications, skill gaps with third-party tools and services present problems for customer organizations.

Customer organizations also need to be on the lookout for data egress strategies when establishing service contracts. Proprietary platforms, or even the way contracts are created and signed, can leave the organization locked into data formats or contract lengths which can jeopardize growth or service improvements. This lock-in condition (Shapiro & Varian, 1999) is problematic for the firm because data migration can be a significant barrier to later growth. Economic models show this kind of lock-in is not always predictable, particularly in markets subject to increasing returns, as technology markets are (Arthur, 1989), and that equilibrium in increasing returns markets can often be suboptimal.

The International Institute for Analytics (IIA), an independent research and advisory firm, found that the biggest barriers to the adoption and effectiveness of BI are difficulty turning insight into action, lack of qualified talent, quantifying value of BI efforts, organizational culture, and a lack of upper management support. The findings result from a 2016 survey of over 300 mid-market and large enterprises across a range of industries (IIA, 2016).

Software developers and service providers enjoy touting the ability to customize their products/services to a client's needs. However, one quickly finds there is a large gap between customization and the ability to configure services using limited options. While configuration options in software applications and services can be quite extensive and may even appear endless, there is an end to a customer's choices. Typically, customization of a software application goes beyond the ability to configure it using developer-provided options and require software (re)engineering to develop extensions to off-the-shelf capabilities.

When BI services are developed or operated on-prem, the firm's ability to completely dictate the capabilities offered by these services is greater. This ability is greatly reduced, at best, when software moves to the cloud, because control over software functionality moves to the vendor and multi-tenancy requires software capabilities to remain consistently available to all customers. Customization remains an option in SaaS environments, but multi-tenancy changes vendor responsibility to its customers (Song, Chauvel, Solberg, Foyn, & Yates, 2017). Configuration, as extensive and beneficial as it may be, is simply not a substitute for customization.

Future Industry Trends

Despite the high levels of interest, importance, and investment in BI in the past decade (Kappelman et al., 2015), service maturity levels remain relatively low. Findings by Forrester Research suggest more time is necessary for providers to develop and converge on industry best practices, incorporate lessons learned, develop technologies and processes suitable to business needs, and determine the right mix of centralized versus distributed offerings (Evelson, 2011).

The growing need for BI-specific technology, services, and data governance at enterprise scale versus the locally optimal solutions prevalent in the market at the moment will continue to drive service maturity. As service complexity needs grow, the initial rush to adoption of disparate tools without an overarching architecture will subside, and more mature architectures will emerge. This has been a long-standing part of the cycle of maturity of corporate IT architecture, which Ross (2003) describes so very well. She identifies four stages of corporate IT architecture with identifying capabilities and increasing levels of service maturity starting with separate, local investment by individual business units without a guiding plan (application silo stage) and proceeding to a controlled list of standardized technologies. Next, a stage she calls rationalized data, which includes the standardization and documentation of not only data formats but also business processes supported by IT, and finally, a stage which enables much greater agility through technology, data, and process modularity brought about by standardization.

Greater architecture maturity and capabilities will also reduce what Gartner calls *technical debt* (Sallam et al., 2017), known elsewhere as lock-in: hasty technical decisions regarding analytics platforms which show quick ROI, become entrenched, and leave the organization stranded or in a more difficult position to grow BI services in the future.

Growth for BI services in the areas of advisory, systems integration, platform selection or development, implementation, and operations will be driven by several trends. These include data governance, tool simplicity, automated knowledge discovery, faster and/or near real-time ETL activities, and mobility. Similar to the development of corporate IT architecture described by Ross (2003), BI services (on-prem and cloud) have begun a process of standardization of capabilities and tools and are evolving from best-in-class applications offering little compatibility with other tools. The eventual standardization of data meaning, aided by mature data governance practices and technologies such as the Extensible Markup Language (XML) and by industry data standards based on XML, will add pressure on BI service providers to compete on tool accessibility via simple interfaces. The increasing power and affordability of IT will also enable a new generation of automatic pattern discovery by BI engines, allowing users to spend less time manipulating data and more extracting actionable insight from it. Simpler insights extend beyond simple descriptive statistics and into significant correlations, meaningful clusters, components of multivalued constructs, etc. Analytical bias is present when existing knowledge influences the conclusions derived from a data set.

ETL activities will also be performed faster and more efficiently, and data access to mobile platforms will be more ubiquitous. Demand for simple, easy-to-use BI platforms which provide greater accessibility to sophisticated data visualizations and which increase business agility will continue to be high. Both of these trends will be enabled by the continued affordability and increased performance of IT components.

Data governance is a subset of IT governance, the management and control of IT planning and implementation to support business objectives (Van Grembergen, 2002). Data governance concerns itself with the technologies, policies, and practices which govern the management of data and other electronic assets (Soares, 2015). Data governance issues include consistent data meaning (semantics) throughout the organization, as well as the more traditional governance concepts of ownership, data operations permissions, etc. Data semantics are an important, but nearly invisible, enabler of BI services. Semantics refers to technologies and processes which enable consistent, well-defined data meaning, allowing systems and people to work with that data based on common understanding. Semantics express portable data and processing rules about the data so that differing terms by separate systems referring to the same concept can be reconciled automatically (Berners-Lee, Hendler, & Lassila, 2001). Data analysis, insight extraction, and automated pattern recognition will all improve as the maturity of semantics policies and technologies matures across industries and within individual firms, fueling growth.

Demand will also rise rapidly for BI services and platforms which can handle the increased speed of multistructured data, i.e., structured data suitable for traditional relational applications, and multimedia, multisource "big data." Trusted data sets are important for all BI activities, and the speed, amount, and variety of big data (McAfee & Brynjolfsson, 2012) streaming into corporations are only accelerating. Lastly, mobility of BI services means more than simply their availability and performance on mobile devices. It will include the ability to provide formats and application links (APIs) so that analysis and results can be inserted into a variety of applications, portals, feeds, and even data-based products (Davenport, 2013).

A couple of factors may also work to slow down full migration to cloud-based services. Data governance practices and/or proprietary data formats may prevent a company from completely moving to a cloud strategy. Data governance practices, formal or not, may make it difficult to overcome the inertia of keeping data "in house." Gartner research (Sallam et al., 2017) shows only 51% of participants in a recent survey about cloud services intend to move to a cloud strategy, an uptick from 46% in 2016, but far from a categorical shift. One must be careful when evaluating such as a result, as intent (as measured in the survey) obscures action, timeframes, percent of existing services being migrated, and even the extent to which a cloud service is truly a cloud effort and not a hybrid solution.

Secondly, work by Forrester (Evelson, 2011) defined untamed business processes as informal, localized, human-dependent, and cross-functional. Untamed business processes create difficulties for organizations looking to standardize BI services (as consumers or providers), because these untamed processes are often not highly visible, and they defy traditional formalization and codification processes and require custom approaches. The research report calls for an agile approach which combines technologies, tools, processes, methodologies, and organizational structures to increase the flexibility with which BI consumers adapt to the changes required by formalized and untamed processes. Not surprisingly, the features of agile BI options in the report include simplicity, automation, consistency, and mobility.

Academic research will continue to be hard-pressed to keep up with these developments, and its largest contribution to this rapidly changing field may be the reinforcement of solid concepts which can accommodate changing technologies while retaining their theoretical validity. Well-understood ideas in project management, customer requirements identification, and data governance, to name a few, combined with new technology developments and emerging data creation and consumption models are key to the formation of BI professionals.

One must remember the cyclical nature of developments in information technology and information systems and that, while BI services are certainly not a fad, the time will come when they are also not front-page news, and new concepts, technologies, tools, and methods will dominate the news cycle. Academic research contributions will continue to add value when theoretical concepts are strong and flexible enough to accommodate these changes.

Bridging the Research-Practice Gap

Researchers in the IT/IS field have made multiple calls for greater technology engagement in academic research (Orlikowski & Iacono, 2001) and to expand how theory is generated to include multiple research perspectives (Orlikowski & Baroudi, 1991), not just the dominant view of most mainstream academic research (Lee, 2010). Other IS academic researchers have observed that academic IS research is in danger of talking about itself, mostly to itself (Keen, 1991), and that three primary effects are to blame: the separation of academia from industry practice, the language and tone of academic research, and the choice of publication channels for the research (Lang, 2003).

This chapter closes with suggestions for how to foster two-way dialogue between academic and industry sectors to develop relevant research which can influence pedagogy and provide the necessary skills not widely found in currently graduates. Partnerships between academe and industry in the IT/IS field are strong but are greatly concentrated to a few programs and researchers. Dialogue can help identify and develop curricular programming or technology infrastructure at universities to help develop skills in business students to define and answer current problems. The suggestions are grouped in three categories: recruiting, pedagogy, and research.

The primary findings of a 2009 survey on the state of university-level BI curricula by the Association for Information Systems were that BI programs were not yet widespread, needed to provide a broader range of skills using interdisciplinary approaches, and require better alignment with industry needs (Wixom et al., 2011). College instructors reported a need for better access to data sets, case studies, textbooks, software, and technical support and training to help them deliver appropriate content. The survey was repeated in 2012 and found demand for skilled BI professionals still outpaces supply despite a tenfold increase in the number of BI degree programs worldwide (Wixom et al., 2014). In the era of Big Data, survey results show foundational skills like communications, business expertise, basic analytics, and data management, and SQL remain the most sought after. Employers participating in the survey express dissatisfaction with the practical experience of graduates. The survey also shows that employers value internships and technical skills like dashboard development and analytics practicums the most (Wixom et al., 2014). Firms can help universities in each of these areas by establishing internship or co-op programs, hosting or sponsoring case study or data analysis competitions, providing financial support for curriculum development, via in-kind donations of hardware/software, or by offering free access to white papers, reports, or data sets for use in classrooms.

Employers report specific tool certification is not a critical element of the skills mix they seek (Wixom et al., 2014), but offering funds or training toward certification for instructors in desirable tools and methods would be useful in curriculum development. Guest lecturers and hosting faculty and students for in-office presentations, networking opportunities, and best practices demonstrations can also spur curricular innovation. Lastly, firms can assist in curriculum improvement efforts by participating in departmental advisory boards.

Academic corporate partners can increase visibility of their recruiting efforts in several ways. Many faculty members appreciate having access to experienced staff who can deliver guest lectures on special topics or skills or who can lead software or analytics demonstrations. This raises the visibility of the company on campuses and generates excitement among students, who see companies whose employees deliver guest lectures as potential sources of employment. Visits to the company's local offices where students learn what the company does and which skills are valued in potential employees are also very useful. The obvious approach of formal internship or co-op partnerships is a tried-and-true method, but too often companies overlook the value and impact of sponsoring student societies and honors societies. Very small investments can lead to tremendous visibility with top-performing student leaders: a few hundred dollars to cover expenses for induction ceremonies, informational meetings, or community service events go a long way on campus. Scholarships and book stipends for students in leadership positions can also greatly enhance the company's prestige in the eyes of future professionals.

Lastly, calls for increased relevance in IS research have been made for decades (Robey & Markus, 1998). Lee (2010) argues that the starting point for IS research does not need to be theory but rather the observation, explanation, and documentation of the *art & craft* of IS professionals. From this, a theory can be derived, but more importantly IS research can document and explain practices as "consumable" for practitioners (Robey & Markus, 1998). Research partnerships between industry and academe already exist but are not as widely accessible as they could be. Funding for work of specific interest to organizations, access to employees or consortium members for the purpose of collecting data for the development of theoretical models, and access to existing corporate data sets would be extremely useful to both academics and corporate partners. Participation in research design and execution, data analysis and synthesis, and publication of results are all valuable commodities to academics and of great use to industry partners. More creative outlets for research cooperation are faculty-in-residence opportunities, in which faculty members can work part time or take a sabbatical to work full time at an organization to develop and deliver research and other products of interest to the organization.

Business intelligence has demonstrated its value to organizations, but as with other IT investment, extracting this value is neither easy nor automatic. Collaboration between academia and industry can result in the improvement of skills for graduates so high in demand, a more prepared professional workforce, and research products which are of more immediate practical use to organizations.

References

Armbrust, M., Fox, A. Griffith, R., Joseph, A.D., Katz, R., Konwinski, A., et al. (2009, February 10). Above the clouds: A Berkeley view of cloud computing. Retrieved from https://www2.eecs.berkeley.edu/Pubs/TechRpts/2009/EECS-2009-28.pdf

Arthur, W. B. (1989). Competing technologies, increasing returns, and lock-in by historical events. *The Economic Journal, 99*, 116.

Barney, J. (1991). Firm resources and sustained competitive advantage. *Journal of Management, 17*, 99–120.

Berners-Lee, T., Hendler, J., & Lassila, O. (2001 May). The semantic web. *Scientific American, 284*(5), 34–43.

Carr, N. G. (2003 May). IT Doesn't matter. *Harvard Business Review*. Retrieved from https://hbr.org/2003/05/it-doesnt-matter

Chaudhuri, S., Dayal, U., & Narasayya, V. (2011). An overview of business intelligence technology. *Communications of the ACM, 54*, 88–98.

Chen, H., Chiang, R. H., & Storey, V. C. (2012 Dec). Business intelligence and analytics: From big data to big impact. *MIS Quarterly, 36*(4), 1165–1188.

Cohen, W. M., & Levinthal, D. A. (1990). Absorptive capacity: A new perspective on learning and innovation. *Administrative Science Quarterly, 35*, 128.

D'Aveni, R. (1994). *Hypercompetition: Managing the dynamics of strategic maneuvering*. New York: Free Press.

Darrow, B. (2015). Shocker! Amazon remains the top dog in cloud by far, but Microsoft, Google make strides. Fortune.com. Retrieved from http://fortune.com/2015/05/19/amazon-tops-in-cloud/

Davenport, T. (2013 Dec). Analytics 3.0. *Harvard Business Review, 91*(12).

Davenport, T. H. (2006). Competing on analytics. *Harvard Business Review, 84*(1), 98.

DeLua, J. (2008). Building a common data foundation for Enterprise Integration and Growth. *Business Intelligence Journal, 13*, 52–56.

Devlin, B. (2010). Beyond business intelligence. *Business Intelligence Journal, 15*, 7–16.

ElBashir, M. Z., Collier, P. A., & Sutton, S. G. (2011). The role of organizational absorptive capacity in strategic use of business intelligence to support integrated management control systems. *The Accounting Review, 86*, 155–184.

Evelson, B. (2011, March 31). Trends 2011 and beyond: Business intelligence. Forrester Research. Retrieved from https://www.forrester.com/report/Trends+2011+And+Beyond+Business+Intelligence/-/E-RES58854

Gartner (2013, February 19). Gartner says worldwide business intelligence software revenue to grow 7 percent in 2013. *Gartner*. Retrieved from https://www.gartner.com/newsroom/id/2340216

Goldman, S. L., Nagel, R. N., & Preiss, K. (1995). *Agile competitors and virtual organizations: Strategies for enriching the customers*. New York: Wiley.

Heizenberg, J., Lo, T., & Chandler, N. (2017, February 14). Magic quadrant for business analytics services, Worldwide. Gartner. Retrieved from https://www.gartner.com/doc/3606022/magic-quadrant-business-analytics-services

International Institute for Analytics (2016). Business intelligence and analytics capabilities report. Portlan, OR (Document No. 108549_G38359.1016).

James, J. (2016, February 14). Magic Quadrant for Business Analytics Services, Worldwide. Gartner. Retrieved from https://www.gartner.com/doc/3606022/magic-quMagic%20Quadrant%20for%20Business%20Analytics%20Servicesadrant-business-analytics-services

Jourdan, Z., Rainer, R. K., & Marshall, T. E. (2008). Business intelligence: An analysis of the literature. *Information Systems Management, 25*, 121–131.

Kappelman, L., Mclean, E., Johnson, V., Gerhart, N., Stewart, B., Peterson, B., et al. (2015). Issues, investments, concerns, and practices of organizations and their IT executives: Results and observations from the 2015 SIM IT trends study. *Society for Information Management*.

Keen, P. G. W. (1991). Relevance and rigor in information systems research: Improving quality, confidence, cohesion and impact. *Information systems research*. Elsevier Science Publishers.

Lang, M. (2003). Communicating academic research findings to IS professionals: An analysis of problems. *Informing Science, 6*, 21–29.

Laskowski, N. (2015). Ten analytics success stories in a nutshell. CIO decisions (CIO.com), published April 27, 2015 [Last accessed April 5, 2018], (at http://searchcio.techtarget.com/opinion/Ten-analytics-success-stories-in-a-nutshell).

Lee, A. S. (2010). Retrospect and prospect: Information systems research in the last and next 25 years. *Journal of Information Technology, 25*, 336–348.

Lee, H., Kim, J., & Kim, J. (2007). Determinants of success for application service provider: An empirical test in small businesses. *International Journal of Human-Computer Studies, 65*, 796–815.

LeMerle, M. (2012). It's time for 'Service as a Service'. *Financial Times*. Retrieved from https://www.ft.com/content/f88bc87a-0e4b-11e2-8b92-00144feabdc0

Liu, A. (2016). TOP 10 Best Cloud Providers 2016. *Cloud Spectator*. Retrieved from https://cloud-spectator.com/best-cloud-providers-2016/

Lowry, P. B., Moody, G. D., Gaskin, J., Galletta, D. F., Sean, H., Barlow, J. B., & Wilson, D. (2013). Evaluating journal quality and the Association for Information Systems (AIS) senior scholars' journal basket via bibliometric measures: Do expert journal assessments add value? *MIS Quarterly, 37*, 993–1012.

Luftman, J., & Derksen, B. (2012). Key issues for IT executives 2012: Doing more with less. *MIS Quarterly Executive, 11*, 207–218.

Luftman, J., Kempaiah, R., & Nash, E. (2005). Key issues of IT executives. *MIS Quarterly Executive., 4*(2), 269–285.

Luftman, J., & Mclean, E. R. (2004). Key issues for IT executives. *MIS Quarterly Executive, 3*, 89–104.

Luftman, J., Zadeh, H. S., Derksen, B., Santana, M., Rigoni, E. H., & Huang, Z. D. (2013). Key information technology and management issues 2012-2013: An international study. *Journal of Information Technology, 28*, 354–366.

Luhn, H. P. (1958). A Business Intelligence System. *IBM Journal of Research & Development, 2*, 314.

McAfee, A., & Brynjolfsson, E. (2012). Big data: The management revolution. *Harvard Business Review, 90*, 59–68.

Mell, P., & Grance, T. (2011). The NIST definition of cloud computing. United States Department of Commerce, National Institute of Standards & Technology (NIST), Report No. 800-145.

Ong, I. L., Siew, P. H., & Wong, S. F. (2011). A Five-Layered Business Intelligence Architecture. *Communications of the International Business Information Management Association (IBIMA.)* Retrieved from http://ibimapublishing.com/articles/CIBIMA/2011/695619/695619.pdf

Orlikowski, W., & Baroudi, J. J. (1991). Studying information technology in organizations: Research approaches and assumptions. *Information Systems Research, 2*, 1–28.

Orlikowski, W., & Iacono, S. (2001). Desperately seeking the IT in IT research: A call to theorizing the IT artifact. *Information Systems Research., 12*, 121–134.

Robey, D., & Markus, M. L. (1998). Beyond rigor and relevance: Producing consumable research about information systems. *Information Resources Management Journal, 11*, 7–15.

Ross, J. W. (2003). Creating a strategic IT architecture competency: Learning in stages. *MIS Quarterly Executive, 2*, 31–43.

Sallam, R. L., Howson, C., Idoine, C. J., Oestreich, T. W., Richardson, J. L., & Tapadinhas, J. (2017). *Magic quadrant for business intelligence and analytics platforms*. Stamford, CT: Gartner.

Sambamurthy, V., Bharadwaj, A., & Grover, V. (2003). Shaping agility through digital options: Reconceptualizing the role of information technology in contemporary firms. *MIS Quarterly, 27*, 237–263.

Shapiro, C., & Varian, H. R. (1999). *Information rules: A strategic guide to the network economy*. Boston: Harvard Business School Press.

Shariat, M., & Hightower, R. (2007). Conceptualizing business intelligence architecture. *Marketing Management Journal, 17*, 40–46.

Soares, S. (2015). *The chief data officer handbook for data governance*. Chicago: Mc Press.

Song, H., Chauvel, F., Solberg, A., Foyn, B. & Yates, T. (2017). How to Support Customisation on SaaS: A Grounded Theory from Customisation Consultants. *IEEE International Conference on Software Engineering*. Buenos Aires, Argentina, IEEE.

Tallon, P. P., Short, J. E., & Harkins, M. W. (2013). The evolution of information governance at Intel. *MIS Quarterly Executive, 12*, 189–198.

Thompson, J. K. (2009). Business intelligence in a SaaS environment. *Business Intelligence Journal, 14*, 50–55.

Turban, E., Sharda, R., Aronson, J. E., & King, D. (2008). *Business intelligence: A managerial approach*. Upper Saddle River, NJ: Prentice Hall.

van der Lans, R. (2009a, March 04). *The flaws of the classic data warehouse architecture, Part 1*. BeyeNetwork. Retrieved from http://www.b-eye-network.com/view/9752

van der Lans, R. (2009b, April 1). *The flaws of the classic data warehouse architecture, Part 2*. BeyeNetwork. Retrieved from http://www.b-eye-network.com/view/9960

van der Lans, R. (2009c, May 12). *The flaws of the classic data warehouse architecture, Part 3*. BeyeNetwork. Retrieved from http://www.b-eye-network.com/view/10342

Van Grembergen, W. (2002). Introduction to the Minitrack: IT Governance and its Mechanisms. *Hawaii International Conference on System Sciences*. Waikoloa, Hawaii.

Vedder, R. G., Vanecek, M. T., Guynes, C. S., & Cappel, J. J. (1999). CEO and CIO perspectives on competitive intelligence. *Communications of the ACM, 42*, 109–116.

Watson, H. J. (2009). Tutorial: Business intelligence - past, present, and future. *Communications of the AIS, 25*, 487–510.

Wixom, B., Ariyachandra, T., Douglas, D., Goul, M., Gupta, B., Iyer, L., et al. (2014). The current state of business intelligence in academia: *The Arrival of Big Data. Communications of the Association for Information Systems, 34*, 1–13.

Wixom, B., Ariyachandra, T., Goul, M., Gray, P., Kulkarni, U., & Phillips-Wren, G. (2011). The current state of business intelligence in academia. *Communications of the Association for Information Systems, 29*, 299–312.

Wixom, B., Yen, B., & Relich, M. (2013). Maximizing value from business analytics. *MIS Quarterly Executive, 12*(2), 111–123.

Yeoh, W., Richards, G., & Wang, S. (2013). Linking BI competency and assimilation through absorptive capacity: A conceptual framework. In *PACIS 2013: Smart, open and social information systems: Proceedings of the 17th Pacific Asia Conference on Information Systems* (pp. 1–9). Association for Information Systems.

Zahra, S. A., & George, G. (2002). Absorptive capacity: A review, reconceptualization, and extension. *Academy of Management Review, 27*, 185.

Chapter 11
Aligning Analytics with Marketing Strategy: Using Analytics to Drive Marketing Strategy with New Media Applications

Lawrence Duke and Atif Ashraf

Abstract This chapter discusses how changes in the business and technological landscape affect marketing practice in industry and academic research. One difference is that marketers have turned to new media and digital marketing tools to understand better their customers and how they interact with their brand. Another development has been the steady convergence between published marketing theory in the academic literature and practitioner research. The chapter also draws attention to how new media marketing and analytics has fostered new insights about the customer journey, such as the creation of the loyalty loop and the need for alignment in marketing strategy. The implications for analytics education are also examined in the chapter with recommendations for curricula shifts and training as they relate to higher demand for and a shortage of qualified graduates. The chapter concludes with a case study, "Einstein Health System: Erectile Dysfunction," which provides a straightforward illustration of the potential benefits for an organization to align their analytical methods with their marketing strategy.

Keywords Marketing strategy · Customer journey · New media marketing · Digital marketing tools · Analytics education · Alignment

Introduction

Marketing strategy continues to evolve in an ever-changing business and technological landscape. Trends tend to change from day to day and in some cases minute by minute. Firms are now required to attend to the particular profile of their consumers and what motivates them to interact with their brand to remain competitive

L. Duke (✉)
Drexel University, Philadelphia, PA, USA
e-mail: lkd26@drexel.edu

A. Ashraf
Sogeti, New York, NY, USA

M. Anandarajan, T. D. Harrison (eds.), *Aligning Business Strategies and Analytics*, Advances in Analytics and Data Science 1,
https://doi.org/10.1007/978-3-319-93299-6_11

in the marketplace. Marketers have increasingly turned to new media applications to help achieve this goal.

New media applications offer an excellent opportunity for firms to capitalize on the data generated by each application. Thoughts and actions of consumers can be analyzed in great detail to provide managers insights into constructing more accurate customer profiles to understand better how to reach out to their clients. Fortunately, recently developed analytical tools help marketers more genuinely grasp the data that is driven by these new media channels. A prime example of a data-driven marketing analysis is assessing the sentiment of a consumer community. By actively monitoring trends within the community, marketers can better understand consumer sentiment through the application of statistical methods. For example, Twitter produces millions of records of unstructured data that can be used to generate a view of consumer sentiment. Analysis of these records can allow marketers to gauge the "pulse" of their consumers. With the Twitter tracking tool Hootsuite, firms can utilize this tool to track the conversations transpiring around their brands using keywords known as hashtags. This data can be processed through Twitter's web API into a big data environment for analysis using methods in artificial intelligence fields such as natural language processing (NLP). Through such processes and practices, managers can better understand their customer feedback and improve the overall brand representation they have in the community. Ultimately, this data-driven analytic approach allows managers to streamline their targeted marketing efforts more efficiently.

The adoption of these new media technologies and the data they generate has created dramatic changes in both industry and the academic world. In academia, curricula have started to shift by introducing these new media applications to the classroom. Theories are being updated to include the social feedback, users becoming a part of the marketing process, and the nature of this new age of marketing. Within the industry, new technologies and techniques are being employed to handle the data generated. Analysts are being hired with both quantitative and technological skills to help make sense of data. The capability of using analytical methods and technology has become a strategic asset for staffing businesses. This challenge has exposed an alignment need both in academia and business. The development of more quantitatively capable marketers requires academia to align with the sector's growing needs for technology and quantitative analysts. As the digital marketing industry itself continues to become more data focused, it is necessary for management to align their business strategy with their analytical insights gained from analysis.

The objective of this paper is to address the evolution of marketing strategy and a better understanding of the customer journey both in academia and the industry, the way analytics enhances traditional academic models, and the issues faced by organizations when trying to align their strategy with analytics. In this context, our paper starts with an overview of the historical progression in marketing theories and the impact of the growth of analytical insights to the marketing discipline in general. Then the paper will address the issues and challenges that businesses and academia currently face in developing alignment. Following that, recommendations will be provided to help organizations enhance their efforts to align their strategies, followed by the coverage of future trends. The paper will conclude with a section covering the implications for marketing strategy formulation and an illustrative case study exemplifying how aligning analytics with digital marketing strategy can yield added value for a firm. This Einstein Healthcare Network: Erectile Dysfunction (ED) case study provides a simple and straightforward example of how analytics and a deeper understanding of the customer journey can help drive bottom-line improvement for a health system's e-mail marketing campaign.

Background

Marketing practices have changed dramatically over the past two decades. Traditional methodologies of reaching the consumers have evolved to incorporating consumers into the marketing process with frequent engagement and the development of feedback loops. We can track the change in the academic coverage of new and innovative marketing practices through the searching of the marketing literature.

Lamberton and Stephen analyzed the research evolution in issues published by academics focused in the area of marketing from the period of 2000–2015 (Lamberton & Stephen, 2016). The two researchers concentrated on the three new media channels of digital, social, and mobile marketing, which they abbreviated as DSMM. Lamberton and Stephen used keyword analysis on 160 published articles in the leading academic marketing journals. Their study identified 200 distinct keywords in the surveyed journal articles and significantly "167 (83.5%) of which were used only once" (Lamberton & Stephen, 2016, p. 148). This finding supports the view that DSMM researchers' work is fragmented (Lamberton & Stephen, 2016).

It is important to point out this vast array of topic areas within digital marketing academic research. New techniques and methodologies have been appearing as consumer sociological trends have shifted to incorporate technological innovations such as social media into people's daily lives. It also brings to light the importance of these technologies as touch points that are critical to a company's marketing success. Lamberton and Stephen (2016) plot out both the keywords found in academic journals, with five or more mentions, and the topics discussed in business publications over the same 2000–2015 period.

Lamberton and Stephen's study points to a dramatic shift toward a more model-based and analytical approach in the published literature (Lamberton & Stephen, 2016). From a business practitioner approach, the change manifests through a new strategic framework development. This framework incorporates a real-time marketing loop process that customers progress through with a greater focus on social media marketing. The academic literature needs to highlight the new marketing theories and structures, as the proper strategic alignment requires the analysis of data acquired through these social media feedback loop platforms. Their work shows that while individual topics mirror themselves from academic research to a business focus, a gap does exist in the attention shown to each subject area. The authors also note, "it is not surprising to see that practitioners have been less focused on the development of analytical methods relative to academics" (Lamberton & Stephen, 2016, pp. 149–150). Yet, some of the more strongly represented topics in the academic world (advertising/search advertising, networks, return on investment [ROI], and user-generated content [UGC]) parallel issues of substantial discussion in the practitioner world (digital advertising, social media and networks, ROI, and UGC), though the ordering of topics diverges to some extent.

More specifically, UGC, one of the most significant research areas among academics and practitioners, is a published content that is "created outside of professional routines and practices" (Kaplan & Haentein, 2010; Wunsch-Vincent & Vickery, 2007; cited in (Smith, Fischer, & Yongjian, 2012, p. 61)). Individuals and groups can produce, modify, share, and consume UGC (Smith et al., 2012). They use social media in many different ways as reflected in their UGC (Kaplan & Haentein, 2010). UGC is related to, but not identical with, electronic word of mouth (eWOM), and while UGC is a broader concept, the two considerably overlap when UGC is brand-related (Smith et al., 2012).

While Lamberton and Stephen focused on the evolution of academic marketing research publications, the development of theoretical marketing models also needs to be discussed. As noted before, social media marketing is a primary topic of interest to business. The social media platforms generate a two-way conversation with consumers. They also facilitate direct interaction with customers and allow real-time analysis of marketing efforts. In turn, these platforms serve as the brand touch points that help drive more profound brand perceptions and broader brand adoption. Consumers can evaluate a brand based on their profile and the UGC they develop on these platforms that serve as an extension of the brand itself. Incorporating the use of these platforms into marketing tactics has caused a fundamental change in underlying marketing theory and strategies.

To help understand the change, we compare the traditional push model from a new one that incorporates the loyalty loop. Court, Elzinga, Mulder, and Vetvik (2009) believe there has been a fundamental shift in the marketing funnel from a conventional push model to one that incorporates the feedback loop. They focused on what they called the loyalty loop, which is centered on developing brand loyalty among consumers. Figure 11.1 illustrates the traditional push model.

The decision-making process is now a circular journey with four phases: initial consideration; active evaluation, or the process of researching potential purchases; closure, when consumers buy brands; and post-purchase, when consumers experience them.

It can be seen from the purchase funnel that consumers first attain a state of awareness then generate familiarity with a brand. This familiarity can occur by examining a direct marketing effort, which creates an impactful association and initial name recognition. They then go on to consider the options or services the brand has to offer. Following that, once their decision has been made, they proceed to make a purchase. Based on the evaluation of their choice, this may result in a loyal customer for an organization.

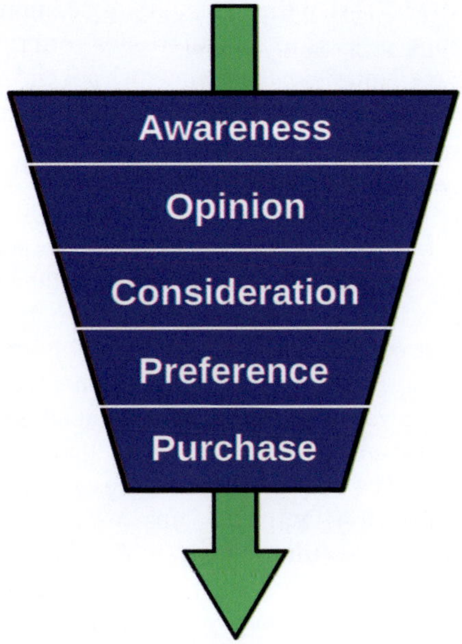

Fig. 11.1 The purchase funnel illustrates the relative number of prospective purchasers over time. (Source: BronHiggs, 2016)

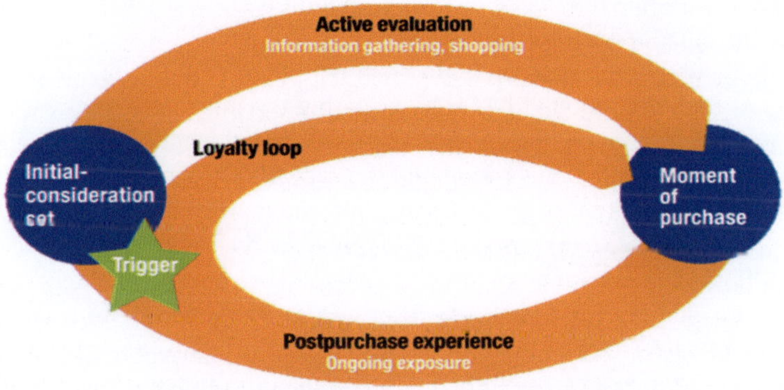

Fig. 11.2 Exhibit from "The consumer decision journey," June 2009, McKinsey & Company, www.mckinsey.com. Copyright (c) 2017 McKinsey & Company. All rights reserved. Reprinted by permission

In today's more digital and technological world, the funnel model has changed significantly to include the numerous touch points that facilitate a two-way conversation with the consumer. Court et al. (2009) visualized the process as shown in Fig. 11.2.

In this version of the loyalty loop, consumers start their initial consideration of a brand from multiple touch points. These contact points can vary from a digital advertisement on a search result, community pages on social media platforms, or a television advertisement. Here the initial impressions are brought directly to the consumer through platforms they engage in on a daily basis. They then move to the second part of the loop, with these impressions already formed, and apply an evaluation process to the companies' offerings. Once they purchase the product, their experience with the brand does not end there. A post-purchase phase then occurs. At this stage, the marketer can continually engage with their customers through soliciting product ratings, follow up e-mails, and feedback on post-purchase surveys.

Employing the loyalty loop model, a marketer must first study where a consumer's first impression occurs. This first step of the consumer journey occurs even before they evaluate the product or brand for a potential choice. The implication for firms is that to reach a consumer, they need to understand who they are and what they are interested in and to engage them to shape their perception. With the multitude of customer touch points, a cohesive and aligned strategy is necessary for firms to target customers effectively. This focused effort to reach out to the client to establish loyalty is aided by a robust analytics capability within an organization. To evaluate the marketer's need for such a skill, it is necessary to understand how analytics enhances traditional marketing models. An evaluation of the purchase funnel (also called an AIDA model) improved by analytics provides this insight.

Elias St. Elmo Lewis, a marketing legend, developed the AIDA model (Lewis, 1908). Marketers use this model (see Fig. 11.1), which stands for awareness, interest, desire, and action, to help manage their advertising campaigns and has proven resilient to the technological changes over time (Hassan, Nadzim, & Shiratuddin, 2015). Even with its resilience, the incorporation of analytics help marketers enhances the value of the traditional models. Using analytics, marketers can track the primary touch points users use to interact with their brand. They can expand the awareness of their content and brand by targeting the most optimal areas shown in the data analysis. The same data also offers marketers the ability to see which content proved to be the most interesting and the critical characteristics of the targeted customer profiles. With these acquired insights, the marketer can target and recommend options to those customers directly to enhance their desire of interacting with the brand. Engaging consumers with personalized and targeted content can potentially raise the chances for a particular user to turn into a conversion. The Einstein Health System case study at the end of this chapter provides a simple example of how analytics enhances the traditional AIDA model through a greater emphasis on the customer journey, touch points, and their importance in both academic research and business practice.

Issues and Challenges

The incorporation of new media marketing platforms has led to a change in the fundamental marketing theories and has enabled organizations to align their strategies through the analysis of the vast amount of customer data now available. With the change comes the inherent need to add insight derived from data analysis to the formulation of marketing strategy. However, for managers to create more efficient and aligned data-driven marketing strategies, organizations need to address several

Fig. 11.3 Company data silos

critical issues. The first issue centers on how an organization starts the alignment process. This action entails the company taking stock of the data the organization already possesses, the data collection methods, and the assessment of potential missing touch points which could provide relevant customer information and data. Here the goal is to understand what role data currently plays in the organization and how the organization stores the data. A frequent issue that occurs in many organizations is the separation of data.

Each department within the company and individual department teams possesses different datasets, and these datasets are located separately from one another with different users accessing them. As you can see in Fig. 11.3, the data is recorded and stored in separate databases, which then form and reconstitute into data analysis assets used by that particular department. In this instance, critical insights that could be used by collating data together are likely lost through the lack of user knowledge between departments. These resulting gaps in developing actionable ideas has become a significant issue that companies face in producing and storing essential documentation around their data and data repositories. Better content management and documentation practices are an easy way for firms to manage and assess their in-house data structure. Understanding how the data is being recorded and mapped, the relevance of the information being collected, and the overview of their data assets can help companies evolve their data strategy. Often when organizations seek to want to understand their data, they do not have the supporting documentation to do so. This lack of supporting documentation requires the individual who has been working with the information the most to provide insights to the company as a whole. This dependence on a particular employee can be a vital issue if that person is no longer a part of the organization and results in a point of failure when developing a data strategy. Organizations need to be aware of the "linchpin" individuals who handle the data in each of their departments and create a central knowledge repository for present-day and future analysts.

The second issue focuses on enhancing data strategy by assessing critical technologies with the goal of improving the skill sets of analysts and streamlining the data flow process. Marketers engage with consumers through a variety of touch points. Each of these touch points serves as a point of data collection. Aggregation of the collected data requires the use of the proper tools. The structure of this data will vary from entirely unstructured to highly structured data. Each data type necessitates that the right systems are in place to store the data and the right tools are available to complete useful analysis. In marketing tactics and strategies, tools are used to enhance the ability of analysts to streamline and leverage search engine optimization results, website analysis for tracking traffic, content marketing platforms for campaign organization, mobile marketing tracking platforms for real-time monitoring, and community management tools for engagement. Given the diversity of functions and needs, many businesses currently set budgets for analytics infrastructure without understanding why a particular analytical application is needed.

These gaps in understanding lead to a broader issue that plaque the digital marketing industry. As digital marketing has become more data focused, there is a significant shortage of analytical industry professionals with the requisite skills to fill the understanding and employment gaps. The profile for a marketing analyst has continued to evolve as the required technology and required skill sets become more complex and varied. An analyst today needs to be able to think critically about the psychology of the consumer through more frequent interaction with technology and a deeper understanding of how to interpret real-time statistics to gain better insights to respond more efficiently to their consumers. Hence, the two issues of focused analytical thinking using data and the adoption of new technologies to attain insights become highly relevant to hiring managers. A number of the essential knowledge skill sets for marketing analysts are highlighted in Fig. 11.4.

A company's collection of more relevant data allows for a better assessment of their strategic marketing effectiveness. However, equally important to the bottom line, companies need to have a sufficient number of trained analysts to dive into the data, conduct proper data management, extract the meaning pieces of data, and analyze the data results. Fully functioning, data-driven marketing departments can capitalize on these opportunities to gain an edge in their marketing strategy by increasing the speed and depth of their analyses. The main issue becomes a supply side one whereby how can industry attract suitably educated professionals to fill these analyst positions.

Business schools can foster the creation of qualified analysts by introducing analytical tools to the classroom and updating the curriculum to include data-driven marketing strategy coverage (Terel & Kapoor, 2016). The marketer of today needs to have a solid grasp of technology, the ability to conduct proper business analysis, and the ability to apply quantitative skills on data to derive actionable insights. While understanding consumer behavior models and theories are necessary,

Fig. 11.4 Analyst knowledge skill sets

marketers now require a more holistic understanding of how the consumer interacts and connects with marketing campaigns. To be more competitive and productive, the marketer needs to leverage the widely used digital marketing analytic tools that can enhance their ability to manage campaigns, aggregate information, conduct analysis, and visualize their data.

In addition to focusing on adding tools to the skills set of aspiring marketers, academia needs to prepare students to be able to think critically about applying proper business analysis and how to effectively present their results. Communication skills are essential in any industry and vital in conducting appropriate data analysis. Understanding how to map a project from beginning to end will allow marketers to understand how their campaign or project interacts at each stage of a process. Proper implementation of these skills gives marketers an edge in seeing potential failures of their plan and allows businesses to manage their limited budgets when developing their data strategy. Learning how to conduct proper business analysis is complex

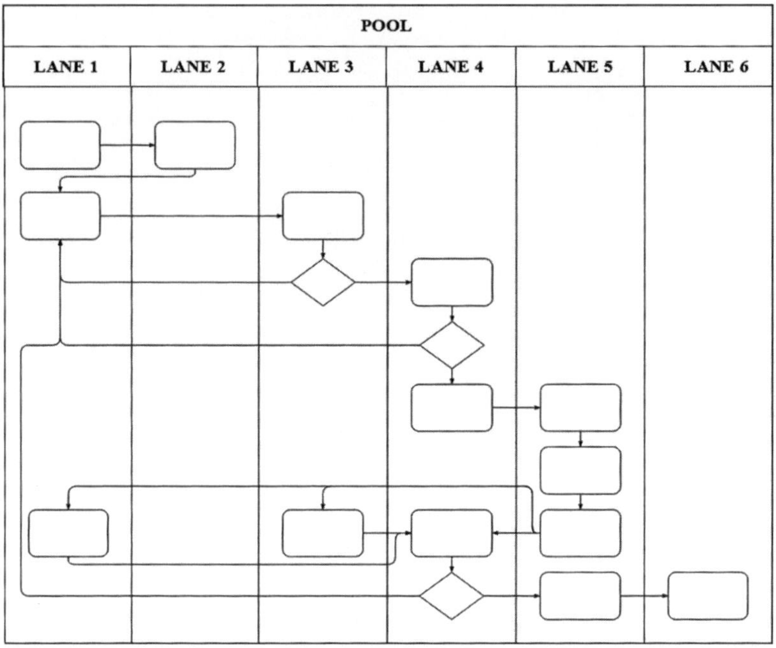

Fig. 11.5 Business process flow diagram

yet can prove invaluable. An example of a tool used in business analysis, the business process flow diagram, can be seen in Fig. 11.5.

Process flow diagram expertise can prove essential when mapping out a campaign, an implementation of a new data flow, or the adding of new data technologies to a company's information systems. A firm foundation of business analysis carries over into every area marketers may operate in and becomes another acquired skill that adds to the productivity and competitiveness of a digital marketer.

The final area that universities need to emphasize in their data analysis courses is visualization of the marketing data. Data visualization supports a more impactful analysis and explicit communication with a target audience. Data visualization enhances dialogue with consumers, which can lead to improved outcomes, including a stronger brand reputation and community engagement. As such, data visualization is a powerful and increasingly essential tool for an analyst. For students, also knowing the correct aspects of design can enhance how their information is received both internally and externally, which can lead to their work having a stronger impact. In fact, companies face an increasing need for individuals who possess these data visualization skills.

The fast-paced nature of digital marketing requires marketers to seek tools that would allow them to increase efficiency, tracking capability, and reach of their content campaigns. This dynamism also requires marketers to think critically and conduct a systematic analysis of information received through their marketing strategies.

These tools will enable marketers to follow the conversation of their consumers on social media, monitor the digital campaign response rates, and provide avenues to conduct more useful competitive research. In fact, thousands of tools are available for marketers with some offered as free for public use. The challenge of individually launching content campaigns can now be reduced to a few clicks of a mouse on one tool. For example, a marketer can follow the performance of a hashtag on Twitter among consumers through Hootsuite®. The use of Hootsuite® and other tools allows marketers to gain a competitive edge and keep current on the results of their campaigns. Moreover, these applications can provide in-depth analytics to marketers enabling them to construct models of performance and effectiveness. These models can then be applied to foster future improvements in advertising and marketing strategies, a critical success factor for marketers facing a more dynamic marketplace. The ultimate challenge for marketers is to gain broader and more in-depth knowledge of these tools and applications and to use those with the best fit to the marketing strategy and objectives.

Solutions and Recommendations

Organizations need to take a step back to evaluate their current marketing and information processes. The most critical action involves a comprehensive marketing audit. Such an investigation would include an inspection of the existing datasets and collected data with a focus on their location, the data source, its accessibility, and any associated documentation needs. The audit outcome will give the organization a standard overview of their data collection processes. Once the processes have been reviewed, it will be necessary to take stock of the current data and to hold an ideation session of the data's analytical potential. Documentation also needs to be addressed in this session as it allows for the sharing of data to be facilitated more easily between the different functions and departments. It also provides a safety net for an organization's information management if their leading analysts leave. The marketing audit offers opportunities to identify additional data collection requirements. This process also helps create a cross-team culture with a focus on data sharing and aggregation. By establishing this comprehensive understanding of who has what data will benefit the organization as a whole. A good analyst will be able to understand how each separate dataset can potentially integrate to drive more significant insights. A central repository would be the best fit, if possible, as it allows for easy access and aggregation.

After the ideation session and the marketing audit have been completed, the organization will have a clearer grasp of its data management. At this stage, firms can assess which analytic tools are truly essential. Marketing analysts should select tools or applications based on more efficient data collection and allow data to be formatted in an analytically useful way. Applications should also be picked based on a better understanding of the data sources and the flows within and through the company. One given tool will likely not solve the overall need of the organization, and dual or multiple application approaches may be required. For instance, an application like Google Analytics can track an e-commerce company's website for usage, while the HubSpot CRM tool provides the capability to monitor their sale conversions and customer data. The resulting data would need to be analyzed, factoring in the primary customer touch points for the e-commerce company. Other important analytical questions would include addressing where do most of the interactions occur and what type of data do they generate. Once these analytical considerations have been factored into the tool and application selection process, it is essential for companies to set up internal training on a selected tool or system. The field of analytics is flooded with many different software vendors, a large number of which are small organizations that may have limited support capabilities. For tools created by smaller organizations, training becomes even more crucial for internal teams to be successful. Furthermore, data analysis can be viewed through various lenses and the use of multiple tools and applications. Our internal biases, analytical training, and personal experiences all shape our analysis. Thus, by marketing management taking such a holistic view, the internal training and employment of multi-trained professionals will likely enhance business opportunities for the firm to gain more valuable insights from data analytics.

Finding talent can be difficult for many organizations. Finding analytical expertise can be even more challenging. However, the study of quantitative analysis and employment in data analysis interest a large and growing number of recent university graduates and current students. For instance, a straightforward way that organizations can capitalize on this interest is to open up an anonymized dataset to the public. As examples, companies can establish an online data competition or host data camps at local universities. Competing or participating in these types of events allows individuals that are interested in the analysis field to engage with your company and increases your known brand recognition as a data forward company. Also, creating an internal training program to teach skills to current employees can prove valuable. It will foster a progressive learning environment and promote a greater corporate brand appreciation among your employees.

In a drive to make their students more attractive to the job market, business schools need to have a curricula shift. This shift requires marketing course content to incorporate the new marketing frameworks, apply critical business analysis, and enhance this analysis through data visualization and other analytical methods. Data has now become a central focus of the marketing world. Marketing graduates increasingly need to have a solid grasp on how to acquire and analyze that data. As an illustration, the curricula shift should incorporate web analytics tools such as Google Analytics, social media tracking instruments such as Hootsuite, statistical tools such as R, and database interaction opportunities. Another focus should continue to be on the psychological aspects of consumer decision-making with an additional emphasis on complementary data analytics using methods such as A/B testing or market basket analysis. Business schools can also provide students with the relevant project management skills. These skills include knowledge of deriving proper requirements for projects, defining project scopes, and conducting project estimation. Also, business schools need to prepare students to understand fundamental design principles and be able to present their data as stories upon entering the workforce. Faculty can also take proactive steps to ensure that their students are proficient in using and applying the current industry tool sets, platforms, technologies, and techniques to real-world problems and case studies. This proficiency involves the creation of a digital ecosystem as part of the curriculum that includes relevant data sources, a "demo" environment, and professional licenses for hands-on experience. Also, many of these tools and technologies allow for professionals to attain certification on their use either through specified courses or self-study. We recommend that business schools take steps to provide for or make available to their students the necessary training to earn these highly sought-after certifications. Students with certifications have a much-increased ability to find desirable positions upon graduation. By augmenting the course of study with the recommendations as mentioned above, business schools can play a more significant role in helping meet growing industry demand for qualified market analysts.

Future Trends

Data collection will continue to abound within marketing to obtain a better understanding of the customer journey; it is the type of touch points that will evolve to provide new and exciting insights. One of the most notable changes that will continue to grow will be a focus on cross-interaction marketing. Cross-interaction marketing is where consumers directly interact with one part of a campaign that leads to another part of the campaign until it reaches its conclusion. We predict that this type of marketing will capitalize on the gamification effect, where users complete tasks or goals to receive rewards for their efforts. These tasks serve as touch points to build customer loyalty and to increase marketing opportunities for new products or services. Coupling with the trend will be a focus on the application of augmented and virtual reality outlets, where users can use the environment around them to interact with brands. Users can discover in-store opportunities, again utilizing the gamification effect, and log this success through apps or limited time web portals. We also see voice and video, along with wearables, as marketing technology growth areas that will generate more touch points and content marketing challenges. Data analysis will continue to play a significant role in gaining greater customer insights and greater alignment in marketing strategy.

Conclusion

The addition and proliferation of new media marketing platforms have prompted a continual evolution of digital marketing. Digital marketing has become a data-centric industry, with technological innovations prompting new touch points that have been integrated into consumer lives. To succeed, organizations need to understand the shift to a more data-focused perspective and prepare their data orientation to handle more massive volumes of data. They face issues stemming from their current data-collecting efforts, digital tool choices, and lack of qualified analysts in the marketplace. In academic research and pedagogical practices at business schools, this shift to a more data-intensive approach needs to be combined with traditional marketing models to include feedback and loyalty loops. Curricula revisions necessitate course development around the myriad of analytical tools used within the industry. Proper alignment of the academic fields with industry practice will lead to a more manageable adoption of an analytical focus in digital marketing strategy and industry practices. When a digital marketing organization plan aligns with their marketing analytics efforts, it leads to valuable insights that will allow the strategy to be more quantitatively defined.

One of the main implications for aligning analytics involves how marketing strategy is conducted. The traditional marketing mix approach that utilizes the 4Ps of product, price, place, and promotion requires refinement to reflect the richer data and information gained from analytics. Palmatier and Sridhar suggest a new framework that incorporates the more open system approach gained from analytics (Palmatier & Sridhar, 2017). This framework includes four fundamental decision-making principles: that customers differ; that customer needs, attributes, and challenges may alter or transform over time; the nature of competitor actions, responses, and game theory elements; and the management of limited human, physical, financial, and informational resources. The open system approach necessitates the use of a number of robust analytical methods and tools, including customer lifetime value, Hidden Markov models, choice models, conjoint analysis, response model attributions, and the like. Through this approach, marketers can better leverage the marketing strategy and mix frameworks through improved measurement and taking into account the interaction between the elements of the four principles.

A number of researchers opine that firms can add value through leveraging analytics (Ducange, Pecori, & Mezzina, 2018; Erevelles, Fukawa, & Swayne, 2016; Kumar & Sharma, 2017; Oztekin, 2017). Their studies provided examples of firms' successful adoption of incorporating analytics in marketing mix components or marketing strategy to improve marketing and business performance. Also, Ducange et al. (2018) argues that data mining and analysis available in social media data has become a requirement for a successful marketing campaign. The Einstein Healthcare case that follows is a suitable example of this argument.

Case Study: Einstein Healthcare Network: Erectile Dysfunction (ED)

Case Study Introduction

The adoption of new media technologies has enhanced the marketer's ability to capitalize on new opportunities. These new opportunities stem from insights derived from data from these new media technologies. The capacity to collect and analyze new media marketing data through the entire marketing cycle can influence the management of the whole flow of a marketing campaign. This capability is particularly beneficial to firms seeking to market specific niche products to customers. It allows brands and sectors of industries, often overlooked due to their specialized audience, to discover the most appropriate customers who want to engage with their product or service.

In the case study that follows, the Einstein marketing team was tasked with promoting their services around the topic of erectile dysfunction (ED) and the services offered by their urology department. ED posed a particular problem to the team as it is layered with inherent cultural sensitivities and perceptions among the public. The team used a multichannel promotional plan to help break through the opinions of ED and to market the expertise of Einstein's urology department. At the end of each promotional marketing step, the team used the data generated to create a marketing funnel of consumers more likely to engage with their services. It can be seen from this case study how the power of data and analytics, managed by a full complement of trained professionals, can foster success for this healthcare company.

Case Study

Einstein Healthcare Network "is a private, not-for-profit organization with several major facilities and many outpatient centers" (Einstein Healthcare Network, 2017a, p. 1). Their urology department has expertise in caring for a patient's urinary tract needs and conditions with a focus on specialized procedures. The urology department approached the marketing team to market their capabilities and to increase enrollment for their specialized surgeries, which earn higher margins than routine care. To better market the services of their urology department, Einstein's marketing team created a wide array of digital marketing content utilized in a multichannel-focused promotional plan.

Often in the initial stages of a marketing plan, marketers have a broad mix of potential consumers to engage with their marketing efforts. It is a real hope that each of those consumers in the initial phase of engagement will turn into an actionable conversion of the product or service offered. This situation isn't often the case, as marketers typically find themselves needing to whittle down the number of consumers into a targeted group which includes those who are most likely to engage with content and has a higher probability of converting.

Einstein's first step was to narrow a number of targeted consumers. In this initial stage, named the "reconnaissance" step, Einstein sent two ED-themed e-mails to the marketing team's entire e-mail list consisting of nearly 156,000 recipients. The team utilized an ongoing monthly e-mail sent to a curated e-mail list to start their efforts. At this point, the group took action by focusing the topic of the e-mail for the month on ED. Following the first e-mail, the team sent a second e-mail with a notification and reminder of an upcoming live chat with a resident doctor from the urology department. These two e-mails were able to generate data in the form of open rates, those who opened or clicked on the e-mail. The monthly e-mail with an ED focus resulted in a 5.2% opening rate, while the reminder e-mail had a rate of 7.7%. Upon seeing the results of each, Einstein used this data to concentrate their target marketing to those who opened the two e-mails. This step-by-step refining process led to a refined, targeted list of 14,263 recipients for the team to further focus their efforts.

With the targeted list in hand, the team sent a follow-up e-mail with a link to the transcript of the live chat with the expert ED urologist. At this stage of the funnel, labeled as the distillation step, the e-mail garnered an outstanding open rate of 44.3%. This result provided evidence of the power of data-driven marketing. By narrowing down their initial list of recipients to those who have shown interest, the team's subsequent efforts would lead to higher engagement and probability of conversion. Again, the users who engaged with this part of the marketing process were moved forward in the funnel to the next stage of marketing efforts.

At this third level of engagement, the targeted communications step, 6088 recipients were targeted. A video featuring the ED expert doctor from the live session was sent to the subscribers resulting in a very high 65.3% open rate. By tracking the user's interests, the team was able to see that these users would continue to engage with the ED subject content. Using an engaged and already proven subscriber of interest and the same ED expert, marketers were able to continue to target and narrow down the number of users. Finally, 3757 recipients were targeted for a final marketing push by the team called the precision marketing step. The expert doctor recorded a video, also included a personal testimonial about his ability that was sent to all the recipients. In this stage, the recipients have repeatedly shown interest, and the team could track them through each step of the process. Again this resulted in a very high open rate of 71.7%. The entire marketing funnel is summarized in Fig. 11.6.

With this targeted data campaign, the team was able to get a return on investment (ROI) of approximately five conversions with a total marketing budget of under $8000. Even though the campaign ROI appears to be a low number, note that each surgery has a high cost resulting in higher revenue to the urology team using a meager marketing budget. With a small marketing investment and the ability to leverage data analytics, a campaign can be managed efficiently with the potential to make a high ROI. Also, the new targeted list was used in conjunction with a now highly

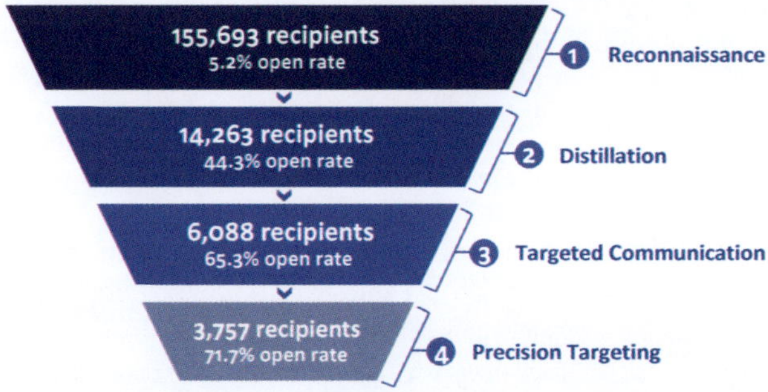

Fig. 11.6 Digital marketing funnel

targeted social media campaign consisting of social ads to the people already identified as good leads through the e-mail campaign.

The Einstein Health System: ED case study is an example of how the power of data-driven digital marketing leads a greater understanding of the customer journey. The Einstein marketing team executed a successful campaign utilizing only a few digital tools. To gather their data and track their efforts, the team used one single analytics platform: HubSpot. HubSpot™ specializes in developing an inbound marketing platform for users to attract their consumers and visitors through digital marketing efforts. For this campaign, the team used the platform to track each e-mail and to create conversion forms to measure the conversion performance of their campaign. Utilizing analytics to drive marketing performance allowed the team at Einstein to engage in precision-targeted marketing that ultimately can lead to success as shown by the attractive ROI and other key performance indicators (Einstein Healthcare Network, 2017b).

References

BronHiggs. (2016). *The purchase funnel (original diagram based on a well-known concept)*. Retrieved from Wikimedia commons website: https://commons.wikimedia.org/wiki/File:The_Purchase_Funnel.jpg

Court, D., Elzinga, D., Mulder, S., & Vetvik, O. J. (2009). The consumer decision journey. *The McKinsey Quarterly, 3*, 96–107.

Ducange, P., Pecori, R., & Mezzina, P. (2018). A glimpse on big data analytics in the framework of marketing strategies. *Soft Computing, 22*, 325–342 https://doi.org/10.1007/s00500-017-2536-4

Einstein Healthcare Network. (2017a). *About*. Retrieved from https://www.einstein.edu/about

Einstein Healthcare Network. (2017b). *Erectile dysfunction workflow copy: Digital marketing funnel*. Retrieved from http://view.ceros.com/einstein-healthcare/ed-work-flow-1/p/1

Erevelles, S., Fukawa, N., & Swayne, L. (2016). Big data consumer analytics and the transformation of marketing. *Journal of Business Research, 69*(2), 897–904 https://doi.org/10.1016/j.jbusres.2015.07.001

Hassan, S., Nadzim, S. Z., & Shiratuddin, N. (2015, January 27). Strategic use of social media for small business based on the AIDA model. *Social and Behavioral Sciences, 172*, 262–269. https://doi.org/10.1016/j.sbspro.2015.01.363

Kaplan, A. M., & Haentein, M. (2010). Users of the world unite!: The challenges and opportunities of social media. *Business Horizons, 53*(1), 59–68.

Kumar, V., & Sharma, A. (2017). Leveraging marketing analytics to improve firm performance: Insights from implementation. *Applied Marketing Analytics, 3*(1), 58–69.

Lamberton, C., & Stephen, A. T. (2016). A thematic exploration of digital, social media, and mobile marketing: Research evolution from 2000 to 2015 and an agenda for future inquiry. *Journal of Marketing, 80*(6), 146–172 https://doi.org/10.1509/jm.15.0415

Lewis, E. S. (1908). *Financial advertising (the history of advertising)*. Indianapolis, IN: Levey Brothers.

Oztekin, A. (2017). Creating a marketing strategy in healthcare industry: A holistic data analytic approach. *Annals of Operations Research*, 1–22. https://doi.org/10.1007/s10479-017-2493-4.

Palmatier, R. W., & Sridhar, S. (2017). *Marketing strategy: Based on first principles and data analytics*. London, UK: Palgrave.

Smith, A. N., Fischer, E., & Yongjian, C. (2012). How does brand-related user-generated content differ across YouTube, Facebook, and Twitter? *Journal of Interactive Marketing, 26*(2), 102–113.

Terel, O., & Kapoor, B. (2016). A business analytics maturity perspective on the gap between business schools and presumed industry needs. *Communications of the Association for Information Systems, 39*. Retrieved from http://aisel.aisnet.org/cais/vol39/iss1/6

Chapter 12
Aligning Data Analytics and Strategy in the Chemical Industry

Suresh Chandran and Rahul Kasat

Abstract Data and analytics are playing a revolutionary role in the chemical industry. This paper provides an overview of the challenges confronting the chemical industry and the opportunities to transform the industry by aligning data analytics and strategy. We look at various facets of the chemical industry and outline the role of data analytics in production and research strategies, as well as in marketing and customer service strategies. Using the case study of DuPont, we provide an example of how applying data and analytics to its precision agricultural technology increased yields and improved productivity. The chemical industry is also successfully implementing analytical techniques used by a variety of other industries such as retailing and finance to create value through differentiation and rethinking customer offerings. We also describe the opportunities that big data and analytics offer the industrial Internet of Things (IoT) strategy to drive performance and growth. Finally, we outline the limitations of data analytics and opportunities for future research in this area and discuss the importance of industry and academia working together to leverage the power of data and analytics in the chemical industry.

Keywords Chemical industry · Data · Analytics · Internet of Things strategy

Introduction

Data and analytics are revolutionizing the chemical industry. Analysis of big data is being used as a tool to assess strategies in manufacturing and pricing and provide insights to evaluate customer demand for innovative products and services. For a chemical company to use big data appropriately, however, it needs to be aligned

S. Chandran (✉)
Drexel University, Philadelphia, PA, USA
e-mail: sc44@drexel.edu

R. Kasat
DuPont, Wilmington, DE, USA

M. Anandarajan, T. D. Harrison (eds.), *Aligning Business Strategies and Analytics*, Advances in Analytics and Data Science 1,
https://doi.org/10.1007/978-3-319-93299-6_12

with and used as a strategic pillar along with people, processes, and technology to create a sustained competitive advantage. Our paper investigates the application of big data to various facets of the chemical industry. To do so, we explore DuPont's experience in aligning analytics with its business strategy.

Background

The definition of big data is based on three Vs: the massive *volume* of data being generated and gathered; the wide *variety* of data types, including text, pictures, and videos from social networking; and the high *velocity* at which the data must be processed. Combined with analytics, the resulting information supports two more Vs: *veracity*, meaning ensuring the quality of the information, and *value* for businesses seeking greater productivity, lower costs, and higher revenues. Big data and analytics have created a paradigm shift in the way chemical companies operate. They have moved from a reactive position in which they respond to changing production or demand conditions to predictive and finally to prescription-based models that determine how best to address these issues in the future (Kaestner, 2016).

Status of the Chemical Industry

The chemical industry is composed of hundreds of companies that manufacture and/or distribute thousands of products such as basic chemicals, synthetic material, agrochemicals, paints and coatings, and inorganic chemicals that are used as intermediates or converted into products for daily use. These products are sold across multiple industries such as oil and gas, pharmaceuticals, automotive, electronics, and packaging. Twenty million people are employed in this industry, which has annual sales of $5 trillion (Statista, 2017), and the industry leaders are large global companies such as BASF, DowDuPont, Sinopec, SABIC, LynodellBasell, Formosa, Bayer, and Mitsubishi Chemicals.

The global chemical industry has been depressed since peaking in 2007. Overall sales growth increased an anemic 2.1% in 2016, as the sector faced declining industrial production and broad inventory rightsizing by many of its customers. The outlook for the industry is expected to be worse for companies that are unwilling to take steps to address the continuing uncertainty in major markets. In the USA, policy changes could make access to international markets difficult. In Europe, monetary easing has not translated into increased demand for chemical companies, and the Brexit cloud affects the outlook in the UK. As the Middle East looks to diversify its economies beyond oil, chemical companies must determine

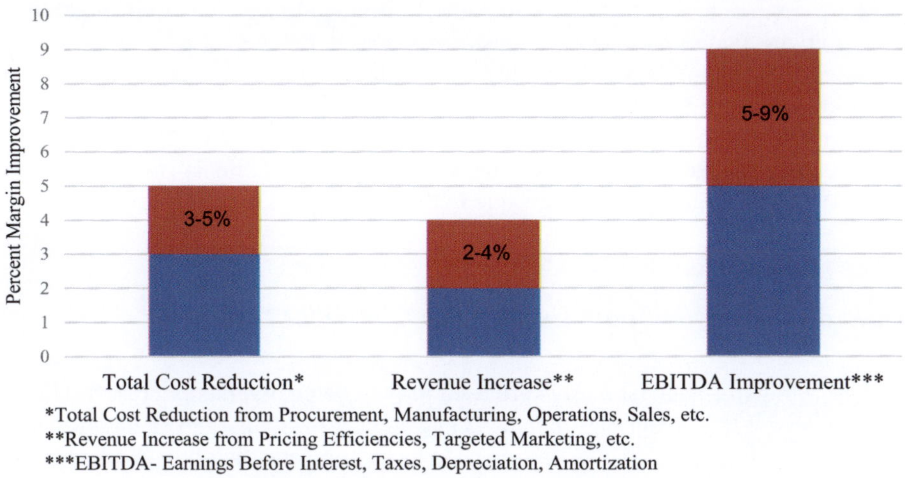

Fig. 12.1 Expected EBITDA margin improvement from digitalization

how best to adapt to this new situation. As the Chinese economy transitions to one based on domestic consumption rather than exports, chemical companies will face further challenges (Sarathy, Morawietz, Gotpagar, & Bebiak, 2017).

Chemical companies are focusing on three strategic activities to address these challenges:

1. Undertaking M&A activity – Faced with challenging demand conditions and profitability, chemical companies are trying to complement their existing offerings or move into lucrative areas aligned with their strategic goals using a mergers and acquisition strategy. In 2016, nearly 1200 deals valued at more than $380 billion were announced including blockbusters such as Dow/DuPont, Syngenta/ChemChina, and Bayer/Monsanto.
2. Emphasizing value over volume – Chemical companies are trying to extract value through differentiation by rethinking customer offerings and service levels with a tiered fee structure and pricing excellence.
3. Adopting digitization and utilizing data and analytics to increase business value – As a recent PWC study (Sarathy et al., 2017) showed, these initiatives can create benefits from an organizational, operational, and customer perspective. Figure 12.1 illustrates that value creation in the chemical industry comes from reducing costs and increasing revenues. Areas ripe for cost reduction include procurement, manufacturing, and operations. From a procurement standpoint, digital procurement planning as well as advanced analytics can lead to cost reductions. From a manufacturing standpoint, robotics, automation, predictive maintenance, and better demand forecasting can lead to cost reduction opportunities. Similarly, from an operations perspective, improving energy efficiencies and performance leads to cost-saving opportunities. In total, utilizing

these approaches can save a company 3–5%. Similarly, enhanced marketing initiatives including targeting customers based on advanced analytics and improving pricing power by having a solutions provider approach rather than a product sales approach can lead to revenue improvements of 2–4%. Furthermore, digitization can result in increases of 5–9% in earnings before interest, tax, depreciation, and amortization. This study found that chemical companies plan to invest 5% of their annual revenue in these initiatives over the next 5 years.

Data and Analytic Tools Used in the Chemical Industry

A wide variety of analytical tools are used in the chemical industry (Briest, Dilda, & Sommers, 2015). Examples include (1) Monte Carlo simulations to identify bottlenecks and the impact of interventions, (2) production and distribution planning to determine product allocations across various product lines given constraints, (3) capacity planning to identify additional capacity requirements given constraints, (4) value-in-use modeling to evaluate the impact of different grades of raw materials across the value chain, (5) optimization of demand and pricing to determine the optimal level of stock keeping units (SKUs) based on their contribution to net profits, and (6) process optimization, meaning improving productivity by optimizing technical parameters in chemical manufacturing.

For example, BASF saved $36 million through analytics using MVT® (Multivariable Testing) analytic software from QualPro. This program utilizes advanced mathematical models that can test up to 40 variables simultaneously. BASF introduced it to optimize its manufacturing and costing as well as improve its product quality at its Freeport facility in Texas. Using this tool, the company was able to identify ways to make improvements in operations as well as optimize its service and marketing strategies (QualPro, 2015). Similarly, Dow Chemical has reported sales and marketing benefits with advanced analytics. Sales forecasts were accurate to within 10%, versus an error rate that was sometimes as high as 40% previously (Kaestner, 2016).

Aligning Analytics and Strategy

Chemical companies have traditionally used data and information technology to reduce expenses. The cost models developed were used to negotiate with suppliers for better terms and reductions in spending (Stringer, 2013). Today, chemical companies use data and analytics to create value and compete effectively in the marketplace. Value creation comes from increasing productivity and efficiency and improving innovation using advanced analytical techniques (Kaestner, 2016). Companies can thus begin to build a competitive advantage by capturing market share and improving margins through data and analytics (Lundia, 2015).

Aligning Data Analytics with Production and Research Strategy in the Chemical Industry

The use of data and analytics in a chemical production plant can help automate production, improve asset utilization, and make better operational decisions in real time. Using data and analytics, energy use, production, and maintenance can be optimized, leading to an improvement in the return on investment. Combining analytics with operational and financial data can streamline supply distribution systems, and models can be developed that lead to better terms when purchasing raw materials and pricing finished goods. Minimizing inventory holding costs through optimization projects enables chemical companies to move products efficiently from manufacturing facilities to numerous destinations at the right time. Analytics software can help enhance demand-planning processes and streamline business operations, enabling chemical companies to reduce costs, increase efficiency, and optimize production (Kaestner, 2016).

For chemical producers with complex operations, production optimization is challenging. Volatility in costs and prices as well as the diversity of inputs is a reality that can be addressed using data and analytics. One global chemical maker used advanced data modeling to deal with all of these problems. Company experts from various disciplines gathered pricing information, equipment information, and market forecasts and input more than 600 decision variables into a mathematical model describing production yields under various operating conditions (Briest et al., 2015). The resulting model helped managers better understand the value chain and highlighted opportunities for improvements that helped increase earnings before interest and taxes (EBIT) returns by more than 50% (Briest et al., 2015). One of the leading advantages of this experience, according to managers, was the realization of better collaboration across different units.

One example of this approach is DuPont's proof of concept for scaling and speeding up the critical comparative genomics workflow in the research environment. Comparative genomics is vital for many research projects, and the number of completely sequenced genomes grows rapidly every year. In this case, the similarities and differences between thousands of genomes, which consist of over four million proteins, had to be evaluated. Traditionally, the process starts with a 2-week initial step of running an all-versus-all comparison of all four million proteins and even running a parallel job on a 1000-core cluster. The results from this comparison are loaded into a relational database that produces a terabyte-sized table – a full 2-day process. This table is then partitioned and indexed accordingly to enable querying within a reasonable time frame. In the last step, a process runs for 30 days, including calling scripts that perform computationally intensive comparisons to find the best bidirectional matches between every pair of genomes and loading the results into the relational database.

To test whether an alternative approach using big data and analytics would prove more efficient, DuPont's bioinformatics experts used Hadoop, MapReduce, and HBase on the Intel cluster to parallel the loading process. They loaded the entire 12-trillion-record data set previously generated by the all-versus-all comparison into HBase. They observed linear scalability as they added more nodes, and the loading process was very fast. They also eliminated the 2-day loading step by using MapReduce to run the all-versus-all comparison and feed the results directly into Hbase as Hadoop generated them. The team was also able to use Hive to immediately query and search the distributed data in HBase, achieving millisecond response times to their queries. As a result, they shrank a 30-day job down to an estimated 4 days with linear scalability (Intel, 2014).

As this case illustrates, the application of faster computational power and more advanced analytic techniques to large data sets is providing continuous improvement in the chemical industry. These advanced analytical applications, some borrowed from risk management and finance, are helping solve previously unsolvable problems such as hidden bottlenecks, operational rigidities, and areas of excessive variability that undermine efficiency in complex manufacturing environments. As companies become better at storing, sharing, integrating, and understanding their data more quickly and easily, they are also seeing improvements in related areas such as quality and production planning (Dhawan, Singh, & Tuteja, 2014).

Aligning Data Analytics with Pricing, Marketing, and Customer Service Strategy

Big data and analytics are now being used to develop competitive pricing strategies based on accurate, timely information from a variety of sources so that more informed decisions can be made. Understanding the distribution of prices and reviewing outliers are key to making adjustments to underperforming products. A leading chemical company in Europe (Mamro, 2014) recently used data and analytics to develop value-based drivers for customer behavior that helped provide specific pricing guidance during contract negotiations. Chemical companies are also using big data to develop accurate forecasts for investment planning purposes (Stringer, 2013).

Case Study of Aligning Data Analytics with Strategy: DuPont

Another area where big data is being used is in precision farming, meaning producing higher yields using fewer chemicals in a very precise way. With precision farming, the detailed analysis of weather, soil conditions, and seed traits helps farmers determine the specifics of planting as well as the application of chemicals to protect their crops. The increasing demand for agrochemicals is an example of how big data is helping create new markets for chemicals (Markets and Markets, 2015).

With increases in population driving the demand for food, more companies are relying on precision agriculture to help farmers make optimal decisions with regard to all aspects of planting, fertilizing, and harvesting crops. Sensors used to measure the temperature and humidity of the soil and surrounding air combined with satellite imagery and robotic drones show crop maturity. When coupled with predictive weather modeling, these images can predict future conditions and help farmers make proactive decisions (IBM, 2015). Advanced precision technology allows farmers access to optimal seed and planting strategies that improve production while minimizing costs and the associated environmental impact. It is projected that global agricultural production in 2050 will be 60% higher than in 2005 as a result of increases in yields and advances in productivity.

One of the leading companies involved in precision farming is DuPont, an American science company founded in 1802 and dedicated to solving challenging global problems using a dynamic portfolio of products, materials, and services to meet the market needs of various industries in more than 90 countries. On August 31, 2017, DuPont merged with the Dow Chemical Company to create the DowDuPont Company, the world's largest chemical company in terms of sales (DowDuPont, 2017). There are several initiatives underway at DuPont to link data analytics and strategy within its operations.

DuPont Pioneer (Pioneer) is the world's leading developer and supplier of advanced plant genetics providing high-quality seeds to farmers around the world. With business operations in more than 90 countries, Pioneer sells high-quality corn, soybeans, sorghum, sunflower, alfalfa, canola, wheat, rice, cotton, pearl millet, and mustard seed, as well as forage additives and a variety of services and expertise. Pioneer combines conventional plant breeding, biotechnology, and emerging areas of science with unsurpassed services and support to deliver solutions to farmers that help them increase agricultural yields to meet the increasing food demands of a growing population (DuPont, n.d.).

Pioneer plays a key role in DuPont's plan to build a higher growth, higher value company, particularly the company's strategic focus on extending its leadership and leveraging links in the science-driven segments of the agriculture-to-food value chains. Pioneer has developed products such as EncircaSM Services that combine the latest technology for weather, soil, agronomy, and analytics to help growers make decisions about farm inputs. Through collaboration with the University of Missouri and the US Department of Agriculture, EncircaSM Services provides growers with a 3D high-resolution view of their soil, showing their depth, texture, organic matter content, and water-holding capacity. These details help growers make more informed decisions about nitrogen and fertility inputs than they could, using publicly available soil maps. EncircaSM Services also helps growers decode the information by providing access to certified service agents, who can help translate data using powerful analytics into specific actions that growers can take to improve efficiency and profitability. In the next decade, EncircaSM Services is estimated to deliver a peak revenue stream of more than $500 million annually for Pioneer (DuPont, 2014). Monsanto, a competitor of DuPont in the agriculture space, recently paid $930 million to acquire the Climate

Corporation, the maker of a software platform that provides weather-related data to help farmers grow crops more effectively. This move highlights the role of analytics in competitive strategies.

DuPont has even empowered the growers' mobile phones and tablets with technological advances. Apps such as PioneerR GrowingPointTM agronomy allow growers to access valuable agronomic information on-the-go. According to the American Farm Bureau Federation, a full 60% of them now use precision data services. These innovative data-driven programs use historical data coupled with new data capture techniques and predictive analytics to help growers maximize crop yields and reduce risk, unlocking the full potential of their land (DuPont, 2016).

Data presented in 2017 by DuPont from 420 Pioneer Research trials conducted in 2015 and 2016 show that the use of data and analytics increased yields by 6 bushels per acre and produced 8 pounds less nitrogen per acre, providing the average farmer with a $22 per acre return (DuPont, 2017).

Emerging and Future Trends

Lessons Learned from Other Industries of Use to the Chemical Industry

Amazon is one of the leaders in predictive analytics and is known for using these tools to target its marketing to increase customer satisfaction, build company loyalty, and increase revenue. Amazon uses big data to analyze what items customers have previously purchased, what is in their online shopping cart or wish list, which products were reviewed and rated, and what items were searched for most. Using this information, the company can recommend additional products that other customers purchased when buying those same items. Amazon's patented anticipatory shipping model uses big data for predicting what products are likely to be purchased, when the buying might occur, and where the products might be needed. The items are sent to a local distribution center or warehouse so they will be ready for shipping once ordered. Amazon uses predictive analytics to increase its product sales and profit margins while reducing its delivery time and overall expenses. Because Amazon wants to fulfill orders quickly, the company links with manufacturers and tracks their inventory. Amazon uses big data systems to choose the warehouse closest to the vendor and/or customer, thereby reducing shipping costs by 10–40%. Additionally, graph theory helps the company decide the best delivery schedule, route, and product groupings to further reduce shipping expenses. Big data is also used to manage Amazon's prices to attract more customers, increasing profits by an average of 25% annually. Prices are set according to activity on the website, competitors' pricing, product availability, item preferences, order history, expected profit margin, and other factors. Product prices typically change every 10 min as big data is updated and analyzed. As a result, Amazon can offer discounts on best-selling items and earn larger profits on less-popular items. For example, the cost of a novel on the *New York Times'* best sellers' list may be 25% less than the retail price, while a novel not on the list costs 10% more than the same book sold by a competitor.

Several other industries including finance, healthcare, and online streaming are using big data to either acquire new customers or retain existing ones. Capital One determines the optimal times to present various offers to clients, thus increasing conversion rates and increasing the ROI from its marketing efforts. Starbucks uses big data about locations, traffic, demographics, and spending behaviors to determine the potential success of each new location.

The chemical industry is learning from techniques used by Amazon and other companies, and customizing it for their own purposes. For example, BASF is deploying a predictive analytics approach by combining the company's historical data with economic data, enabling it to forecast demand. The forecasting model considers external factors such as weather, macroeconomic data, political, legal, and regulatory changes and internal factors such as BASF's proposed strategies to make decisions. Using the forecasting model, BASF can adapt its plant as demand changes. BASF is also using advanced analytics models for predictive asset management, plant commissioning, and process control. For example, the company automated the production of liquid soaps at its pilot plant in Kaiserslautern, Germany. Once a user places an order for a customized soap in this smart pilot plant, the radio-frequency identification tags attached to the soap containers update the production equipment of the desired soap specifications and enable mass customization (Deloitte, 2016).

Industrial Analytics and the Internet of Things for Chemicals: Top Priorities to Build a Successful Strategy for the Future

The intelligent systems used in the chemical industry have the potential to transform the industry by improving operations and providing the technology that enables the capture of process data that can be quickly converted into effective actions. The Internet of Things (IoT) helps connect assets, people, products, and services to enable real-time decision-making while reducing costs and boosting innovation. Some uses of big data and analytics in the IoT strategy to drive performance and growth are outlined below (SAP, 2016):

- Predictive Maintenance – Software and analytics embedded in a chemical company's assets make them more intelligent and help diagnose their health. Assets can send signals about their status and performance to predict possible malfunctions and maintenance needs.
- Precision Farming – For precision farming to work effectively, all participants in this ecosystem – farmers, agribusiness suppliers, equipment manufacturers,

traders, and technology providers – have to be synchronized. A big data-enabled platform is needed to share the information that supports precision farming.

- Operational Intelligence – Chemical firms generate a great deal of data during manufacturing but use only a small amount to improve decision-making. By using big data and analytics in real time, performance can be improved through sound decision-making.
- Smart Products and Connected Logistics – Chemical firms can use sensors and active RFID tags to identify the specifics of products such as their location and authenticity. This capability helps track products and ensures their quality across the value chain. Supply chain risks can be minimized by using devices that create and process data in real time.

Case Study of Industrial Analytics and IoT Strategy: GE

GE is one of the key companies in the industrial sector that has invested heavily in data analytics in recent years. The company has been installing sensors in turbines, jet engines, and machines so it can collect data and analyze it in real time to help improve the productivity and reliability of these machines, resulting in savings of billions of dollars. GE sees about a $1 billion opportunity in the oil and gas industry. A single unproductive day on a platform can cost a liquified natural gas facility as much as $25 million. Increasing the uptime of these operations is critical, especially given that oil and gas companies are facing a drop in revenues due to lower crude prices. GE believes that software, data, and analytics will be central to the company's ability to differentiate itself within the oil and gas industry. Therefore, it began developing a cloud-based software platform called Predix that provides real-time data to operators and maintenance workers to reduce machine downtime. GE believes Predix can help the oil and gas industry address four of its most pressing challenges: improving asset productivity, creating a real-time picture of the status of an entire operation, stemming the costly loss of tacit knowledge from an aging workforce, and building an industrial Internet platform that meets customer needs (Winnig, 2016).

Challenges and Limitations of Data Analytics: Opportunities for Future Research

The big data movement is creating opportunities for the chemical process industries to improve their operations. However, there are challenges as well.

Challenges and Opportunities with Regard to Volume

Data collected from industrial processes reflect their operation under normal conditions and production situations. The range of this data is relatively narrow even though the quantity of information may be huge. Such data sets have limitations for predictive activities and control and optimization tasks. To optimize processes, a design of experiment program is needed to actively collect data over a wide range of conditions. Future research should focus on the development of methods to remove irrelevant data, as well as analytical methods including visualization tools that can condense large amounts of information (Reis, Braatz, & Chiang, 2016).

Challenges and Opportunities with Regard to Value

Chemical industry data come from different sources and involve analysis of a variety of objects and are therefore complex. In addition to the usual scalar quantities such as temperature and flow, data collected in modem industrial settings also include other data structures such as one-way, two-way, and higher-order arrays indexed by time and/or space that are representative of a product or process. Future research should focus on developing more platforms that can fuse all of these heterogeneous sources of information together (Reis et al., 2016).

Challenges and Opportunities with Regard to Veracity

When dealing with the analysis of big data, the quality of the information is of utmost concern. Quality issues include how the data are collected, whether the information is current, and whether there is uncertainty in measurement. Improving processes requires an assessment of the multiple sources of variability. Reducing process variation and increasing product quality and consistency are the basis of most improvement activity in this regard. Ensuring that data are acquired correctly and using appropriate statistical tools are essential to the analysis of big data sets in the future (Reis et al., 2016).

Challenges and Opportunities with Regard to Velocity

When collecting large quantities of data at high speed in chemical plants, several challenges are inherent, including appropriate collection techniques and resolution. Future research should develop ways to select the proper resolution that takes the features of the variables into account (Reis et al., 2016).

In the chemical industry IoT world, the use of information will become more sophisticated. Given that various players in the value chain capture the data used, strategic partnerships and business models that enable value creation through collecting, analyzing, and making sense of the data have the potential to transform the industry.

Bridging the Gap Between Theory and Practice

To achieve growth in the chemical industry, and to address the challenges and opportunities of volume, variety, veracity, and velocity described above, it is important that industry and academia work together to develop tools and skill sets to leverage the power of big data and analytics. The American Institute of Chemical Engineers (AIChE) has started to organize sessions focused on big data and analytics for the chemical industry during its annual and topical conferences. In addition, in terms of workforce development, it is important that universities develop an interdisciplinary curriculum combining elements of traditional chemical engineering education along with big data analytics. Academia can also provide on-the-job training for experienced professionals within the industry and offer professional courses for the chemical industry on the use of big data analytics. It is also important that chemical companies partner with companies such as Amazon, Facebook, Alphabet, and Microsoft that have developed expertise in the area of data analytics to leverage already developed tools. Academia can play a crucial role in building consortia with different stakeholders to facilitate this partnership.

Conclusion

Companies that consult in the area of strategy believe that many organizations have not yet recognized how big data and analytics can transform their company's ability to operate efficiently and innovate (KPMG, 2015). Chemical companies should focus on the following action areas as they consider their big data and analytics strategies (Kaestner, 2016):

- Focus on the business issues – Understanding the issues will provide clues to where big data capabilities can have the greatest impact.
- Insights do not come from data alone – Data must be analyzed and then applied to the issues that the business needs to address.
- Go beyond immediate solutions – Companies must focus on the enterprise as a whole when using data and big analytics solutions.
- Clearly articulate the value proposition – The source of competitive advantage, whether cost or differentiation, must be articulated, and data and analytics need to be linked to this value.
- Understand that customers drive decisions – Companies must use data and analytics to engage in profitable transactions with their customers.

- Analyze the right data – Companies must understand the business problem and the data needed for a solution to it.
- Share success stories – Companies must share stories and knowledge about problem-solving across the organization so that more projects can be funded.
- Focus on stakeholders – Companies must engage with all stakeholders and communicate and demonstrate the power of data and analytics to transform the organization.

About 2.5 quintillion bytes of data are generated every single day, and 90% of the world's data have been created in the last 2 years alone (IBM, 2017). As the availability of data grows, it offers companies greater opportunities to focus on the customer while increasing efficiency and revenue. To realize this promise, companies will need to delve deeper into unstructured data and leverage advanced analytics and computing to generate valuable insights.

As chemical companies shift from a pay-by-the ton revenue model to provide value-added products and services model, they will need to plan and identify how to integrate their digital and physical assets across different stages of the value chain. To operate and grow their businesses, they will need to use data and analytics to increase the effectiveness of their operations while also thinking about innovative ways to grow their business by discovering new and smarter materials and creating new service-driven value propositions (Deloitte, 2016).

It is important for academics and students to recognize that successful chemical companies use big data effectively by dealing with it in a strategic manner. In other words, they identify a particular business problem and use big data and analytics as part of the solution with the goal of creating value. The identified value proposition is therefore important and allows for solutions, be they lower costs, differentiation, or reduction of risk, to be achieved using big data. Advanced analytics, therefore, is a tool to implement an answer that drives value. Leaders within a company must be able to clearly articulate its purpose and then translate it into action throughout the organization. As competitive dynamics within an industry evolve, insights unleashed by analytics should be used to improve competitive performance and alignment, focus on stakeholders, and achieve strategic business objectives.

References

Briest, P., Dilda, V., & Sommers, K. (2015). Taming manufacturing complexity with advanced analytics. *McKinsey Quarterly*. Retrieved from http://www.mckinsey.com/business-functions/operations/our-insights/taming-manufacturing-complexity-with-advanced-analytics

Deloitte. (2016). *Deloitte insights: Industry 4.0 and the chemical industry*. Retrieved from https://dupress.deloitte.com/dup-us-en/focus/industry-4-0/chemicals-industry-value-chain.html#endnote-sup-3

Dhawan, R., Singh, K., & Tuteja, A. (2014). When big data goes lean. *McKinsey Quarterly, 24*(2), 97–105. Retrieved from http://www.mckinsey.com/business-functions/operations/our-insights/when-big-data-goes-lean

DowDupont. (2017). *Global leaders in agriculture, materials science and specialty products*. Retrieved from http://s21.q4cdn.com/813101928/files/doc_downloads/resources/DowDupont-New-Fact-Sheet.pdf

DuPont. (2014, March 12). DuPont leader outlines growth strategy for seed business. *DuPont Media Center*. Retrieved from http://www.dupont.com/corporate-functions/media-center/press-releases/dupont-leader-outlines-growth-strategy-for-seed-business.html

DuPont. (2016, July 11). Big data helps growers produce more. *DuPont Media Center*. Retrieved from http://www.dupont.com/corporate-functions/media-center/featured-stories/july-2016/big-data-agriculture.html

DuPont. (2017, March 1). *Global agriculture and chemicals conference*. Retrieved from https://s2.q4cdn.com/752917794/files/doc_presentations/2017/mar/DuPont-Slides-BAML-Conf-Mar-1-2017.pdf

DuPont. (n.d.). *DuPont Pioneer – improving farmer productivity around the world*. Retrieved from http://www.dupont.com/corporate-functions/our-company/businesses/pioneer.html

IBM. (2015). Using predictive weather analytics to feed future generations. *IBM Research*. Retrieved from http://www.research.ibm.com/articles/precision_agriculture.shtml; https://intrepidsr.wordpress.com/2015/08/30/precision-agriculture-using-predictive-weather-analytics-to-feed-future-generations/

IBM. (2017). What is big data? *IBM InfoSphere Platform*. Retrieved from https://www-01.ibm.com/software/data/bigdata/what-is-big-data.html; https://www.ibm.com/blogs/insights-on-business/consumer-products/2-5-quintillion-bytes-of-data-created-every-day-how-does-cpg-retail-manage-it/

Intel. (2014). *DuPont Proof of Concept Explores Hadoop Environment for Big Data and Big Science*. Retrieved from https://pdfs.semanticscholar.org/8df2/3487cc2679dcbd8a67ee26d60b9ae572ca77.pdf

Kaestner, M. (2016). Big data means big opportunities for chemical companies. *KPMG REACTION Magazine – Twentieth Edition*. Retrieved from kpmg.com/reaction

KPMG. (2015). Going beyond the data: Turning insights into value. *KPMG White Paper*. Retrieved from https://assets.kpmg.com/content/dam/kpmg/pdf/2015/08/going-beyond-the-data-turning.pdf

Lundia, S. R. (2015, May 26). How big data is influencing chemical manufacturing. *Chem Info*. Retrieved from https://www.chem.info/blog/2015/05/how-big-data-influencing-chemical-manufacturing

Mamro, S. (2014, March 26). Marketing executives learn to leverage big data for pricing and profitability. *Fourth Source*. Retrieved from http://www.fourthsource.com/data/marketing-executives-learning-leverage-big-data-pricing-profitability-18764

Markets and Markets. (2015, July 13). Agricultural adjuvants market accelerating rapidly. *Precision Farming Dealer.* Retrieved from https://www.precisionfarmingdealer.com/articles/1542-agricultural-adjuvants-market-accelerating-rapidly

Qualpro. (2015). BASF case study – chemical industry process improvement. *Qualpro White Paper*. Retrieved from https://qualproinc.com/case-studies/basf/; https://qualproinc.com/case-studies/basf/

Reis, M. S., Braatz, R. D., & Chiang, L. H. (2016). Big data challenges and future research directions. *Chemical Engineering Progress, 112*(3), 46–50.

SAP. (2016). *CEO perspective: The internet of things for chemicals*. Retrieved from http://www.iotsworldcongress.com/wp-content/uploads/2016/01/document.pdf

Sarathy, V., Morawietz, M., Gotpagar, J., & Bebiak, J. (2017). 2017 Chemical industry trends. *Strategy&*. Retrieved from https://www.strategyand.pwc.com/trend/2017-chemicals-industry-trends

Statista. (2017). *Revenue of the chemical industry worldwide 2002–2016 (in billion dollars)*. Retrieved from https://www.statista.com/statistics/302081/revenue-of-global-chemical-industry/

Stringer, B. (2013, December 13). Big data and the chemical industry. *ICIS*. Retrieved from https://www.icis.com/resources/news/2013/12/13/9735874/big-data-and-the-chemical-industry/

Winnig, L. (2016, February 18). GE's big bet on data and analytics. *MIT Sloan Management Review, 57*(3). Retrieved from https://sloanreview.mit.edu/case-study/ge-big-bet-on-data-and-analytics/

Index

M. Anandarajan, T. D. Harrison (eds.), *Aligning Business Strategies and Analytics*, Advances in Analytics and Data Science 1,
https://doi.org/10.1007/978-3-319-93299-6

Printed by Amazon Italia Logistica S.r.l.
Torrazza Piemonte (TO), Italy

28071890R00132